SpringerBriefs in Earth Sciences

For further volumes:
http://www.springer.com/series/8897

Harsh K. Gupta · Vineet K. Gahalaut

Three Great Tsunamis: Lisbon (1755), Sumatra–Andaman (2004) and Japan (2011)

Springer

Harsh K. Gupta
National Disaster Management Authority
New Delhi
India

Vineet K. Gahalaut
CSIR, National Geophysical Research Institute
Hyderabad
India

ISSN 2191-5369 ISSN 2191-5377 (electronic)
ISBN 978-94-007-6575-7 ISBN 978-94-007-6576-4 (eBook)
DOI 10.1007/978-94-007-6576-4
Springer Dordrecht Heidelberg New York London

Library of Congress Control Number: 2013934100

Printed on acid-free paper

Springer is part of Springer Science+Business Media (www.springer.com)

Preface

The idea of doing a SpringerBrief on the "Three Great Tsunamis" emerged while talking with Petra van Steenbergen at the Springer Stall at Taipei on August 9, 2011 during the 8th Annual Convention of the Asia Oceania Geosciences Society (AOGS). I (the first author) was at the Springer Stall in connection with the "Author Meet and Greet" event for the just published "Encyclopedia of Solid Earth Geophysics" edited by me. The Encyclopedia was very well received. During our conversation Petra told me about the SpringerBriefs and how these briefs focus on a topical issue in a simple, easy to be understood by all approach. We had organized a special session during the 8th AOGS Convention dedicated to the scientific work on the Great Mw 9.0 Tohoku earthquake and the resultant tsunami that devastated Fukushima and the nearby regions a few months earlier on March 11, 2011. On December 26, 2004, the mega Sumatra–Andaman earthquake of Mw 9.1 had occurred, which had claimed an estimated 2,30,000 human lives. I was deeply involved with setting up of India's Tsunami Warning System, which had become operational in September 2007. I was also invited to a symposium in Portugal to revisit the November 1, 1755 great Lisbon earthquake and the resultant tsunami that had claimed close to 1,00,000 human lives and had totally destroyed Lisbon; and the tsunami related developments over the past two and a half centuries. Earlier, I had also listened to a very interesting talk by Prof. Carl Fuchs where he had compared the happenings at Lisbon during 1755 and in east and South–East Asia during the 2004 Mw 9.1 earthquake generated tsunami. I talked with Petra about these three tsunamis. She liked the idea very much and encouraged me to consider writing about the three tsunamis as a SpringerBrief book. I kept toying with the idea of doing a SpringerBrief on the Three Great Tsunamis, however, time was a constrain. I met Petra a few months later during the Fall Meeting of the American Geophysical Union at San Francisco in December 2011, and Petra suggested that I take a co-author to expedite the writing of the book. I discussed about this project with Vineet Gahalaut, a colleague of mine, and he graciously agreed to be a co-author.

With the passage of time the impact of earthquakes and the resultant tsunamis is on an increase. This is in spite of tremendous developments in natural and social sciences in the past few decades. It is interesting to note that at the time of writing the 'Preface', just 12 years of the twenty-first century have been completed. However, in these 12 years, more human lives are lost due to earthquakes and the resultant tsunamis than the entire twentieth century. Since the occurrence of the Lisbon tsunami in 1755, a lot has happened. The Lisbon earthquake is credited to be beginning of the science of Seismology. At that time, there was a great debate whether the earthquake was an act of God or a natural phenomenon. The questionnaire sent then to collect the information about the earthquake has been a great source of information and research even today. By the time of the 2004 Sumatra–Andaman tsunami, the science of locating earthquakes, and issuing tsunami warnings was well developed. However, tsunamis in the Indian Ocean had been very rare. In the twentieth century, only three tsunamis, not very large ones, had occurred in the Indian Ocean. The huge number of lives lost was partly due to ignorance: many people walked into the bare ocean floor as the tsunami trough had emptied the near coast sea floor. Moreover, December 26, 2004, being a Sunday and the tsunami occurred in the morning hours and there were many taking a morning walk near the sea shore and got killed. Another reason of high number of lives lost and the immense loss of property was the flouting of the laws which prohibit commercial activity within 500 m from the high tide line in most countries. The situation was totally different for the March 11, 2011 Tohoku earthquake of Mw 9.0. Japan is the country which is most frequented tsunamis. However, the magnitude of the 11 March earthquake far exceeded the estimated size of earthquakes in that region, and the tsunami defense measures were falling short of the requirements for the occasion. It must, however, be mentioned that but for the scientific, technological, and administrative measures taken in Japan, the loss of human lives and property would have been much more.

It is hoped that this book would provide an interesting reading to many desirous to learn about tsunamis, and developing a tsunami resilient society.

We would like to thank Petra van Steenbergen for support to this project of writing. Several of our colleagues at the National Disaster Management Authority (NDMA), National Geophysical Research Institute (NGRI) and Indian National Centre for Ocean Information and Services (INCOIS) helped in compiling the text for this book. Springer needs to be thanked for timely and beautiful production of this book.

Harsh K. Gupta
Vineet K. Gahalaut

Contents

Introduction

Tsunamis are primarily caused by earthquakes. Under favorable geological conditions, when a large earthquake occurs below the sea bed and the resultant rupture causes a vertical displacement of the ocean bed, the entire column of water above it is displaced, causing a tsunami. In the ocean, tsunamis do not reach great heights but can travel at velocities of up to 1,000 km/h. As a tsunami reaches shallow sea depths, there is a decrease in its velocity and an increase in its height. Tsunamis are known to have reached heights of several tens of meters and inundate several kilometres inland from the shore. Tsunamis can also be caused by displacement of substantial amounts of water by landslides, volcanic eruptions, glacier calving, and rarely by meteorite impacts and nuclear tests in the ocean.

In this SpringerBrief, the causes of tsunamis, their intensity and magnitude scales, global distribution and a list of major tsunamis, are provided. The three great tsunamis of 1755, 2004, and 2011 are presented in detail. The 1755 tsunami caused by the Lisbon earthquake, now estimated to range from Mw 8.5 to 9.0, was the most damaging tsunami ever in the Atlantic Ocean. It claimed an estimated 1,00,000 human lives and caused wide-spread damage. The 2004 Sumatra–Andaman Mw 9.1 earthquake and the resultant tsunami were the deadliest ever to hit the globe, claiming over 2,30,000 human lives and causing wide-spread financial losses in several South and South–East Asian countries. The 2011 Mw 9.0 Tohoku-Oki earthquake and the resultant tsunami were a surprise to the seismologists in Japan and around the globe. The height of the tsunami far exceeded the estimated heights. It claimed about 20,000 human lives. The tsunami also caused nuclear accidents. This earthquake has given rise to a global debate on how to estimate the maximum size of an earthquake in a given region and the safety of nuclear power plants in coastal regions. This Brief also includes a description of key components of tsunami warning centers, progress in deploying tsunami watch and warning facilities globally, tsunami advisories and their communication, and the way forward.

Chapter 1
Fundamentals of Tsunamis

Abstract This chapter deals with the causes, characteristics, occurrences, propagation, intensity and magnitude scales etc. of tsunamis. A worldwide list of major tsunamis is provided.

1.1 Introduction

The word tsunami (pronounced as tsoo-nah-mee) is a Japanese word that means *harbor wave (tsunami)*. This is a word that has been coined by Japanese fishermen. They would return from the sea to find that their villages had been destroyed by large waves. While at sea, they did not see or experience waves large enough to wash away a village. There are very few other languages that have an equivalent native word for tsunami. In the Tamil language (Tamilnadu, an east coast province of India), the word is *aazhi peralai (destruction big-waves)*. In the Acehnese language (in Aceh, Sumatra, Indonesia), it is *iё beuna* or *alôn buluёl.* In the Defayan and Sigulai languages, spoken on Simeulue Island, off the western coast of Sumatra in Indonesia, the word is *smong and emong*, respectively.

Although several instances of historical tsunamis have been reported, based on the identification of paleotsunami deposits and their dating, the most specific historical record of a tsunami is from the Malian Gulf (a gulf of the Aegean Sea). In the summer of 426 BC, a tsunami hit the Malian Gulf between the northwest tip of Euboea and Lamia. The Greek historian Thucydides described how a series of earthquakes during 431–404 BC occurred causing a tsunami that affected the region. Remarkably, he could correlate the earthquake and the tsunami. The Roman historian Ammianus Marcellinus described a typical sequence of a tsunami, including an incipient earthquake, the sudden retreat of the sea followed by a gigantic wave during the AD 365 tsunami that devastated Alexandria. Japan is the nation with the most recorded tsunamis in the world. The earliest recorded disaster being the AD 684 Kakuho earthquake. The number of significant tsunamis in Japan totals 195 since AD 684, averaging one event every 6.7 years, which is the highest rate of occurrence

H. K. Gupta and V. K. Gahalaut, *Three Great Tsunamis: Lisbon (1755), Sumatra–Andaman (2004) and Japan (2011)*, SpringerBriefs in Earth Sciences,
DOI: 10.1007/978-94-007-6576-4_1,

in the world. These waves have hit with such violent fury that entire towns have been destroyed. The latest being the 2011 tsunami due to the Tohoku earthquake.

In this chapter, we will deal with the causes of tsunamis, where they occur, their characteristics and scales to measure the intensity and magnitude of tsunamis. At the end of the chapter, we provide a worldwide list of major tsunamis.

1.2 Causes of Tsunamis

Earthquakes produce about 90 % of tsunamis. Most of the earthquakes (~90 %) are of tectonic origin. These earthquakes occur due to a sudden release of elastic energy along a fault plane. A fault can be defined as a fracture plane, along which there has been a displacement of the two sides of the fracture plane relative to one another. These faults can be divided into two main categories, viz., strike-slip and dip-slip faults. A strike-slip fault is a vertical plane where the two sides of the plane move horizontally relative to each other along the strike of the fault (Fig. 1.1). Depending upon the sense of motion (clockwise and anticlockwise) these faults are further divided into right-lateral (sinistral) or left-lateral (dextral) strike-slip faults. To simplify this concept, imagine yourself standing on a fault, with the fault line going in-between your feet. When the block on the right side of the fault moves towards you, it is referred to as a right-lateral strike-slip fault. In the case of a dip-slip fault, the motion is along the dip of the plane. A dip-slip fault can either be a normal or reverse fault (Fig. 1.1). A normal fault is inclined and the rocks above it move downward relative to each other. A reverse fault is also inclined but the rocks above it move upward relative to those below it. A reverse fault with a gentle dip is known as a thrust fault. A combination of a strike-slip fault and a dip-slip fault is known as an oblique fault. These faults exist everywhere on the Earth. However, they are abundant at plate margins and a slip along them produces earthquakes. According to the plate tectonic theory, it is considered that the Earth is made up of several plates

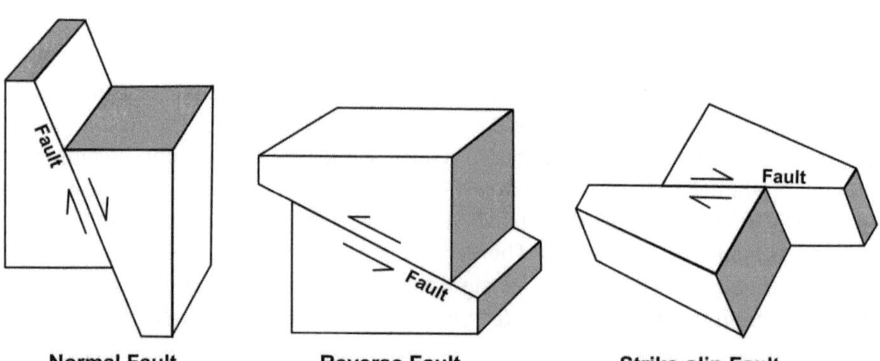

Fig. 1.1 Faults in the Earth's crust

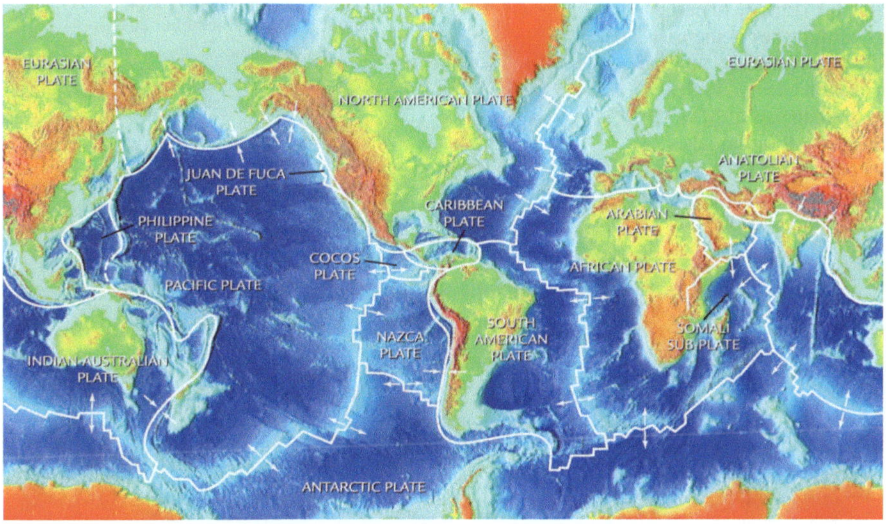

Fig. 1.2 Major plates of the Earth. The majority of the earthquakes occur at plate boundaries. The boundaries with the *diverging arrows* (divergent plate margin) do not produce large magnitude earthquakes. However, the convergent plate margins (shown with *converging arrows*) produce large magnitude earthquakes

that are in continuous motion with respect to each other (Fig. 1.2). These plates, with a thickness of about 100 km, deform at the boundaries and hence a majority of earthquakes occur there. Large magnitude earthquakes generally occur at convergent plate margins where two or more plates collide or subduct. These regions produce large magnitude earthquakes (M > 7) and account for 96 % of the strain energy release. The divergent plate margins at mid-oceanic ridges, where new crust is continuously created, rarely produce large magnitude earthquakes.

Here, we are discussing tsunamis generated by three mega earthquakes of magnitude ≥9. Below is the annual frequency of earthquake occurrence globally.

Descriptor	Magnitude	Annual frequency
Great	8 and higher	1
Major	7–7.9	18
Strong	6–6.9	120
Moderate	5–5.9	800
Light	4–4.9	6,200
Minor	3–3.9	49,000

The energy released by a 6 magnitude earthquake is similar to the energy of the Hiroshima kind of nuclear bomb. The energy release increases with the magnitude. A magnitude 7 earthquake releases 30 times more energy than a magnitude 6 earthquake and so on. So a magnitude 9 earthquake is equivalent to 2700 magnitude 6 earthquakes. M ≥ 9 earthquakes occur very infrequently. It is estimated that

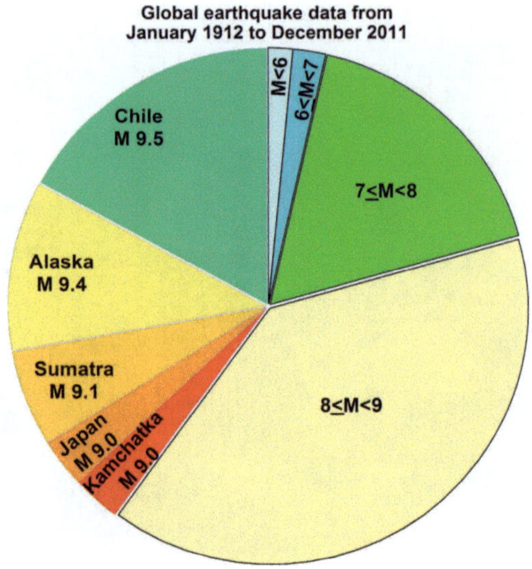

Fig. 1.3 Energy released by earthquakes in the past 100 years globally. 96 % of the strain energy is released by earthquakes of M > 7. Five M ≥ 9 earthquakes accounted for 36 % of the energy in the past 100 years (*Data source* USGS)

1–3 M 9 earthquakes may occur per century (McCaffery 2008). It is interesting to note that earthquakes of M > 8 account for a major portion of the seismic energy released globally (Fig. 1.3).

A tsunami can be generated when the sea floor abruptly deforms and vertically displaces the overlying water column (Fig. 1.4). This can happen due to the occurrence of an earthquake, volcanic eruption and a landslide. Almost all tsunamis are caused by earthquakes. A tsunami can be generated when a shallow focus (focal depth < 70 km) large magnitude (M > 6.5) earthquake occurs on a fault and the rupture causes a vertical displacement in the ocean bottom, resulting in the displacement of the overlying column of water. It is therefore found that the earthquake generated tsunamis are mostly due to reverse and normal fault dominated earthquakes.

Tsunamis can also be caused by an undersea landslide. The landslide may occur by slope failure, which can be triggered by earthquake shaking. Underwater landslides that generate tsunamis are called sciorrucks. Sciorrucks rapidly displace a large volume of water causing a tsunami. Their existence was confirmed in 1958, when a giant landslide in Lituya Bay, Alaska, caused the highest tsunami wave ever recorded, the height being 524 meters. Extremely large volcanic eruptions and landslides produced by sciorrucks can generate tsunamis that travel far distances. Historical examples of surprise landslide tsunamis include the huge tsunami generated by the 1929 Grand Banks and the 1998 Papua New Guinea earthquakes.

Rarely can a tsunami be generated from surface detonations, explosive volcanoes or asteroid strikes. Asteroids with diameters >200 m, capable of generating a tsunami, impact the Earth about once every 10,000 years. Thus, although there exists the possibility of a tsunami generation due to asteroid impact, it is extremely rare.

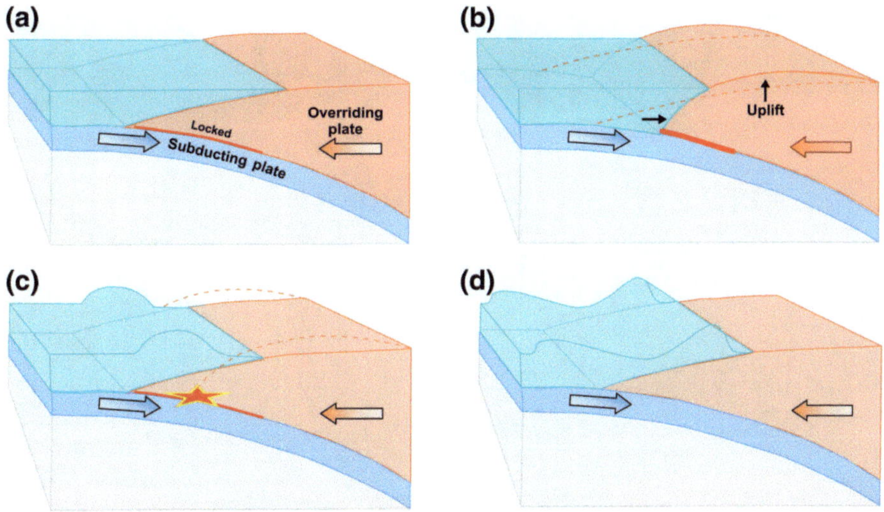

Fig. 1.4 Tsunami generation due to an earthquake. The first two panels show the strain accumulation at a subduction zone (**a, b**). The occurrence of an earthquake (star in c) causes a sudden motion that disturbs the water column above it and leads to the generation of a tsunami (fourth panel) (**c, d**)

1.3 Where They Occur

Almost 80 % of the world's tsunamis occur in the Pacific Ocean (Fig. 1.5). The subduction zones around the Pacific Ocean, known as the Ring of Fire, produce large magnitude earthquakes that cause tsunamis. The Ring of Fire starts from New Zealand, heading northwest to Indonesia, west to Papua New Guinea and Indonesia, northeast along the Asian coastline, east to North America and then south along the western North American coastline. The other major region of tsunami generation is the Java Sumatra subduction zone. Tsunamis have also been reported by earthquakes in other parts of the world, e.g., the Makaran region (off the coast of Pakistan and Iran), the Mediterranean Sea and part of the sea off the coast of Portugal.

1.4 Tsunami Propagation and Characteristics

Tsunami waves are gravity-driven water waves. They belong to the same family as common sea waves but they are distinct in their mode of generation and in their physical traits. Unlike common sea waves (tide, wind wave, surges, etc.) that evolve from persistent winds, tidal effects, etc., in most cases tsunamis are generated by a sudden vertical movement of the ocean floor. These sudden changes can occur from undersea earthquakes, landslides and volcanoes. The wind driven swell might have a period of about 5–10 s and a wavelength (λ) of up to 150 m. A tsunami, on the other

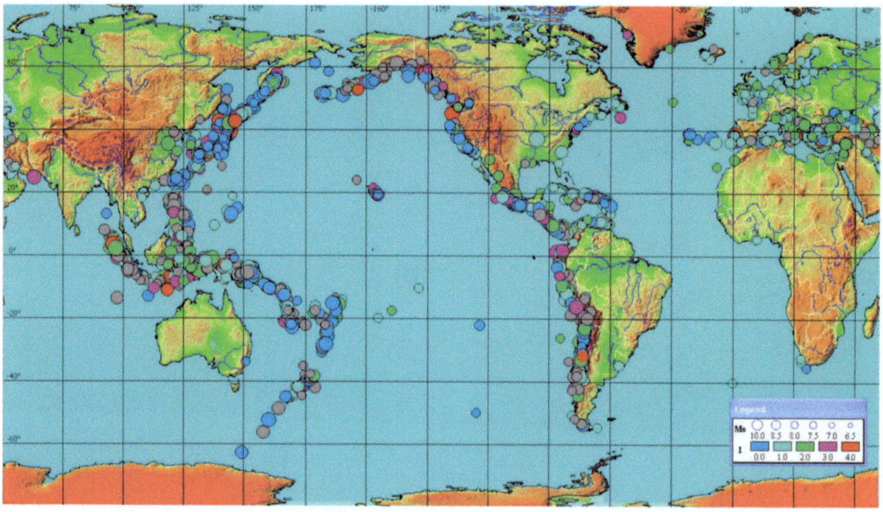

Fig. 1.5 Global historical tsunamis (for the period from 1628 BC to 2011). The size of the circles is proportional to the event magnitude with color representing the tsunami intensity on the Soloviev-Imamura scale. *Source* Global Tsunami Database (GTDB), Tsunami Laboratory, ICMMG SD RAS, Novosibirsk, Russia

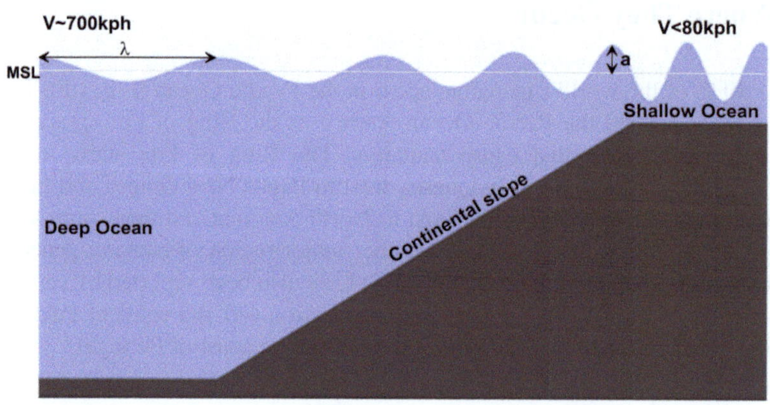

Fig. 1.6 Propagation of a tsunami. As it propagates towards the shore, its wavelength (λ) and velocity (V) decrease, whereas the amplitude (a) increases since the frequency remains the same. Increase in amplitude is referred to as 'shoaling'

hand, can have a wavelength in excess of 100 km and period of 1 min to 1 h. From a hydrodynamics point of view, these waves are long ($\lambda >> H$, H being the depth of the ocean). The propagation velocity, c, may be determined from the formula, $c = \sqrt{g}H$, where g is the acceleration due to gravity. For $H = 4$ km (typical ocean depth), the tsunami waves propagate with a velocity of about 720 km/h (Fig. 1.6). Tsunamis, with their longer period and higher velocity, have much longer wavelengths than beach waves. In the deep ocean, a tsunami spans 10, 30 even 100 km between crests. With wavelengths this large, tsunami slopes are very small even if the wave has large amplitude. For ships at sea, the tsunami passes completely unnoticed. Tsunamis have a small amplitude (wave height of about 30 cm) offshore and a very long wavelength (often hundreds of kilometers long, whereas normal ocean waves have a wavelength of only 30 or 40 m), which is why they generally pass unnoticed at sea, forming only a slight swell (20–30 cm for large tsunamis) above the normal sea-surface. They grow in height when they reach shallow water. As a result of their long wavelengths, tsunamis behave as shallow-water waves. A wave becomes a shallow-water wave when the ratio between the water depth and its wavelength is small. Because the rate at which a wave loses its energy is inversely proportional to its wavelength, tsunamis not only propagate at high speeds but they can also travel great, transoceanic distances with limited energy loss. In oceans with a uniform depth, tsunamis propagate out from their source in circular rings with ray paths that look like spokes on a wheel. As the velocity of a tsunami depends on depth, its propagation is sensitive to the shape of the sea floor. Under water ridges and irregularities of the sea floor cause a scattering of tsunamis. Consequently, the tsunami's travel time and amplitude have to be adjusted according to the prevailing ocean floor conditions and bathymetry. Another important characteristic that differentiates tsunamis from other waves is that all other waves lose their amplitude with depth. As a matter of fact the sea is very quiet at a depth of approximately 2 km, whereas with a tsunami the entire column of water from the sea-surface to the ocean bottom is in motion.

Shoaling. Towards the shore, oceans become shallow and the waves carried by them amplify in a process called **shoaling**. The tsunami velocity depends on the ocean depth so as water shallows, tsunami waves slow down. Because their frequency is fixed, the wavelength of a slowing tsunami decreases and since the energy remains the same, the amplitude increases. Secondly, because a tsunami occupies the entire water column, as it enters shallow water its energy also becomes compressed vertically. The only way for the compressing wave to maintain the same energy flux is for it to grow in amplitude.

Run-Up. Run-up is the final phase of a tsunami's life. The run-up phase begins when the approaching tsunami shoals to an amplitude roughly equal to the water depth and the wave begins to break (Fig. 1.7). Run-up also covers the inundation phase where the water runs over land and reaches its maximum excursion above sea level.

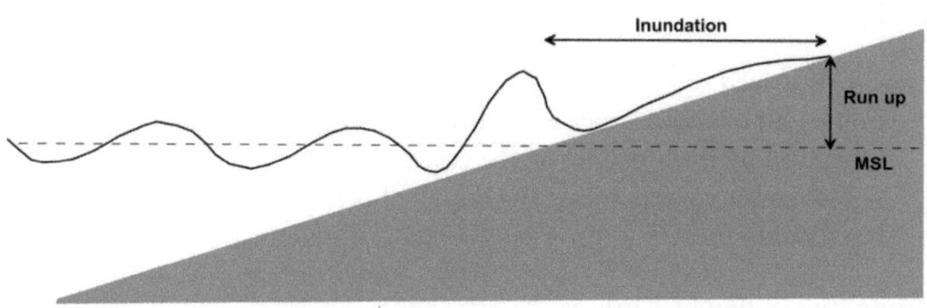

Fig. 1.7 Inundation and run up height of a tsunami

Inundation. This is the horizontal extent of the tsunami on land (Fig. 1.7).

Drawback. Drawback (a 'drawing back' of the water) is a phenomenon in which the ocean recedes before a tsunami strikes a coast. If the first part of a tsunami to reach land is a trough, rather than a crest of a wave, it is called a drawback. The water along the shoreline recedes dramatically, exposing normally submerged areas. A drawback can exceed hundreds of meters and people unaware of the danger sometimes remain near the shore, or even walk into the emptied sea floor to satisfy their curiosity or to collect fish from the exposed sea-bed and are drowned when the tsunami crest arrives.

1.5 Measuring the Tsunami

1.5.1 Intensity Scales

The first scales routinely used to measure the intensity of a tsunami were the *Sieberg-Ambraseys scale*, used in the Mediterranean Sea and the *Imamura-Iida intensity scale*, used in the Pacific Ocean. These scales were similar to the earthquake intensity scales where the tsunamis were described on a six-point scale from very light (1) to disastrous (6). These have been replaced by a 12-point scale, proposed in 2001 by Gerassimos Papadopoulos and Fumihiko Imamura. The tsunami scale is arranged according to (a) the tsunami's effects on humans, (b) effects on objects including boats and (c) damage to buildings. The scale is as follows.

I. Not felt.

II. Scarcely felt.
   a. Felt by a few people onboard small vessels. Not observed on the coast.
   b. No effect.
   c. No damage.

III. Weak.
   a. Felt by most people onboard small vessels. Observed by a few people on the coast.
   b. No effect.
   c. No damage.

IV. Largely observed.
   a. Felt by all onboard small vessels and by a few people onboard large vessels. Observed by most people on the coast.
   b. A few small vessels move slightly onshore.
   c. No damage.

V. Strong. (wave height 1 m)
   a. Felt by all onboard large vessels and observed by all on the coast. Few people are frightened and run to higher ground.

b. Many small vessels move notably onshore and a few of them crash into each other or overturn. Traces of a sand layer are left behind on the ground with favorable circumstances. Limited flooding of cultivated land.
c. Limited flooding of outdoor facilities (such as gardens) of near-shore structures.

VI. Slightly damaging. (2 m)
a. Many people are frightened and run to higher ground.
b. Most small vessels move violently onshore, crash strongly into each other, or overturn.
c. Damage and flooding in a few wooden structures. Most masonry buildings withstand.

VII. Damaging. (4 m)
a. Many people are frightened and try to run to higher ground.
b. Many small vessels are damaged. A few large vessels oscillate violently. Objects of variable size and stability overturn and drift. A sand layer and accumulations of pebbles are left behind. A few aquaculture rafts are washed away.
c. Many wooden structures are damaged and a few are demolished or washed away. Grade 1 damage and flooding in a few masonry buildings.

VIII. Heavily damaging. (4 m)
a. All people escape to higher ground and a few are washed away.
b. Most of the small vessels are damaged, many are washed away. A few large vessels are moved ashore or crash into each other. Big objects drift away. Erosion and littering of the beach. Extensive flooding. Slight damage in tsunami-control forests and stop drifts. Many aquaculture rafts are washed away and a few are partially damaged.
c. Most wooden structures are washed away or demolished. Grade 2 damage in a few masonry buildings. Most reinforced-concrete buildings sustain damage and in a few, grade 1 damage and flooding are observed.

IX. Destructive. (8 m)
a. Many people are washed away.
b. Most small vessels are destroyed or washed away. Many large vessels are moved violently ashore and a few are destroyed. Extensive erosion and littering of the beach. Local ground subsidence. Partial destruction in tsunami-control forests and stop drifts. Most aquaculture rafts are washed away and many partially damaged.
c. Grade 3 damage in many masonry buildings and a few reinforced-concrete buildings suffer from grade 2 damage.

X. Very destructive. (8 m)
a. General panic. Most people are washed away.
b. Most large vessels are moved violently ashore and many are destroyed or collide with buildings. Small boulders from the sea-bottom are moved inland. Cars are overturned and drifted. Oil spills and fires start. Extensive ground subsidence.

c. Grade 4 damage in many masonry buildings and a few reinforced-concrete buildings suffer from grade 3 damage. Artificial embankments collapse and port breakwaters are damaged.

XI. Devastating. (16 m)

a. Lifelines are interrupted. Extensive fires. Water backwash drifts cars and other objects into the sea. Big boulders from the sea-bottom are moved inland.
b. Grade 5 damage in many masonry buildings. A few reinforced-concrete buildings suffer from grade 4 damage and many suffer from grade 3 damage.

XII. Completely devastating. (32 m)

a. Practically all masonry buildings are demolished. Most reinforced-concrete buildings suffer from at least grade 3 damage.

There is another scale that directly connects the tsunami height with the intensity.

$$I = \frac{1}{2} + \log_2 H_{av}$$

where H_{av} is the average wave height along the nearest coast. This scale, known as the *Soloviev-Imamura tsunami intensity scale*, is used in global tsunami catalogues.

1.5.2 Magnitude Scales

The first scale to calculate the magnitude of a tsunami was the ML scale, ML = 2 (logE-19), proposed by Murty and Loomis (1980). It is based on the potential energy, E, of the tsunami wave. Difficulties in calculating the potential energy of a tsunami mean that this scale is rarely used. Abe introduced the *tsunami magnitude scale* M_t, as follows,

$$M_t = a \log h + b \log \Delta + D$$

where h is the maximum tsunami-wave amplitude (in m) measured by a tide gauge at a distance Δ (in degrees, one degree = 110 km) from the epicenter and *a*, *b and D* are constants.

1.6 List of Major Tsunamis

Table 1.1 is a list of major global tsunamis compiled from various sources (modified from Wikipedia).

Table 1.1 List of major tsunamis

Place	Date	Earthquake M	Height (m)	Death (due to ground shaking and the tsunami)	Remarks
Pacific coast of Japan	March 11, 2011	9.0	10–30	20,000	Multiple hydrogen explosions and a nuclear meltdown at the Fukushima I nuclear power plant
South Island, New Zealand	February 22, 2011	6.3	3.5	181	Aftershock of the 2010 earth-quake; 30 million tons of ice tumbled off the Tasman Glacier into Tasman Lake. Many buildings collapsed lor were significantly damaged
Chile	February 27, 2010	8.8	1.8–9.0	550	Very strong shaking and aftershocks damaged fisheries
Samoa	September 29, 2009	8.1	4.6–6.1	189	Main street flooded, cars overturned, shoreline business was affected and the water system also got damaged
Niigata, Japan	July 16, 2007	6.8	0.2–0.5	11 deaths and 1000 injuries	Buildings were completely destroyed
Solomon Islands	April 24, 2007	8.1	12	62 deaths and more than 5000 residents were displaced	The largest waves completely destroyed the two villages, Tapurai and Riquru and beaches shifted outwards of up to 70 m
Kuril Islands	November 15, 2006	8.3	15	$10 million in damage to the docks	A larger tsunami wave, following earlier small ones, crossed the Pacific and damaged the harbor at Crescent City, CA, USA

(continued)

Table. 1.1 (contiuned)

Place	Date	Earthquake M	Height (m)	Death (due to ground shaking and the tsunami)	Remarks
South of Java Island	July 17, 2006	7.7	2–6	800	The earthquake occurred as a result of thrust-faulting on the boundary between the Australian and Sunda plates
Indian Ocean	December 26, 2004	9.1–9.3	33	230,210	The largest earthquake generated a tsunami and the longest duration of faulting (~10 min) severely affected the coastal regions of Indonesia, Thailand, India, Sri Lanka and Somalia
Papua New Guinea	July 17, 1998	7.1	15	2,200	Completely destroyed several villages
Okushiri, Hokkaido, Japan	July 12, 1993	7.8	2.6–32	230	Large landslide
Sea of Japan	May 26, 1983	7.7	10	107	Damaged the fishing harbor of Wajima
Spirit Lake, Washington, USA	May 18, 1980	Volcano	Eruption column rose 24,400		Caused a mega-tsunami with highly toxic water with volcanic gases seeping up from the lake bed, raising the surface elevation of the lake by over 60 m. Also lake water was displaced 800 feet up the hillside
Tumaco, Colombia	December 12, 1979	7.9		259 dead, 798 wounded and 95 missing or presumed dead	Destruction of six fishing villages and the city of Tumaco
Moro Gulf, Mindanao, Philippines	August 16, 1976	7.9		5,000 dead, 2,200 missing or presumed dead	Devastated the cities of Cotabato, Pagadian and Zamboanga

(continued)

Table. 1.1 (contiuned)

Place	Date	Earthquake M	Height (m)	Death (due to ground shaking and the tsunami)	Remarks
Alaska, USA	March 27, 1964	9.4	30	143	Produced earthquake liquefaction and several landslides
Niigata, Japan	June 16, 1964	7.5–7.6	Moderate tsunami	28	Wide spread liquefaction and apartment buildings were destroyed in the port of Niigata city
Vajont Dam, Monte Toc, Italy	October 9, 1963	250-m high mega-tsunami wave		1,450	Flooding destroyed the villages of Longarone, Pirago, Rivalta, Villanova and Fae
Valdivia, Chile	May 22, 1960	9.5	25	6,000	Caused a volcanic eruption, one of the most destructive tsunamis of the twentieth century
Lituya Bay, Alaska, USA	July 9, 1958	7.9	Washing 524 m	5	Highest recorded mega-tsunami. The Lituya subglacial lake dropped 30 m and produced a giant 524 m wave
Severo-Kurilsk, Kuril Islands, USSR	November 5, 1952	9.0	15–18	Out of a population of 6,000 people, 2,336 died	Destruction of many settlements in Sakhalin Oblast and Kamchatka Oblast, the main impact struck the town of Severo-Kurilsk
Aleutian Islands	April 1, 1946	7.8	13–40	165	Multiple destructive waves obliterated the Scotch Cap Lighthouse on Unimak Island, Alaska
Nankaidō, Japan	December 21, 1946	8.4	5–6	Tsunami washed away 1451 houses and caused 1500 deaths	Destroyed 36,000 homes in southern Honshu

(continued)

Table. 1.1 (contiuned)

Place	Date	Earthquake M	Height (m)	Death (due to ground shaking and the tsunami)	Remarks
Tonankai, Japan	December 7, 1944	8.0	10	1223	26,146 houses were destroyed and 47,000 houses were seriously damaged
Showa Sanriku, Japan	March 2, 1933	8.4	12–15	Lost 42 % of its total population and 98 % of its buildings. Destroyed about 5,000 homes and killed 3,068 people	Three hours after the main shock, a magnitude 6.8 aftershock, followed by 76 more aftershocks (with a magnitude of 5.0 or greater) occurred over a period of six months
Newfoundland	November 18, 1929	7.2	7	28	Large tsunami waves snapped telegraph cables laid under the Atlantic
Kanto, Japan	September 1, 1923	7.9	12	142,800	Duration of the earthquake was between 4 and 10 min, followed by 57 aftershocks. Fire spread due to the high winds of a typhoon (over 570,000 homes were destroyed, leaving an estimated 1.9 million homeless). Most lives were lost due to ground shaking and fire
Messina, Italy	December 28, 1908	7.2	12	100,000–200,000	Large undersea landslide due to normal faulting
Meiji Sanriku, Japan	June 15, 1896	7.2	30	27,000	The tsunami was caused by slope failure triggered by the earthquake and the rupture velocity was unusually low due to an accretionary wedge

(continued)

Table. 1.1 (contiuned)

Place	Date	Earthquake M	Height (m)	Death (due to ground shaking and the tsunami)	Remarks
Krakatoa, Sunda Strait, Indonesia	August 26–27, 1883		40	120,000	Caused pyroclastic flows, volcanic ashes and tsunamis. A volcanic winter reduced temperatures worldwide by an average of 1.2 °C for the next 5 years and recorded the largest explosion
Arica, Chile	August 16, 1868	8.5		70,000	Three military vessels anchored at Arica were swept away by the tsunami
Hawaiian Islands	April 2, 1868	8.0	18	31	Caused a landslide on the slopes of the Mauna Loa volcano
Edo, Japan	November 11, 1855	7.0		7,000	Earthquake was followed by 78 aftershocks in the first month
Nankai, Tokai, and Kyushu Japan	December 4–7, 1854	3 quakes in 3 days, two had magnitudes of 8.4 and 7.4	4–8.4	80,000–100,000	Destructive deep thrust quake
Sumatra, Indonesia	November 25, 1833	8.8–9.2		No reliable records of the loss of life (unknown)	Earthquake shaking lasted 5 min in Bengkulu and 3 min in Padang
Mount Unzen, Nagasaki Prefecture, Kyūshū, Japan	May 21, 1792	6.4	100	15,000	The worst volcanic disaster. Caused a landslide and an earthquake due to the collapse of domes
Yaeyama Islands, Okinawa, Japan	April 4, 1771	7.4	40–80	12,000	The population decreased to about one third of what it was before the earthquake and agriculture was severely damaged

(continued)

Table. 1.1 (contiuned)

Place	Date	Earthquake M	Height (m)	Death (due to ground shaking and the tsunami)	Remarks
Lisbon, Portugal	November 1, 1755	8.5–9.0	15	60,000–100,000	The earthquake caused gigantic fissures (5 m wide) and the tsunami traveled with a speed of 400 km/h
W. Hokkaido, Japan	August 29, 1741			1,467	The tsunami associated with the eruption of the volcano on Oshima island caused a large landslide
Hōei, Japan	October 28, 1707	8.4	10–20 m	30,000	Hot springs at Yunomine, Sanji, Ryujin, Seto-Kanayana and Dogo stopped and more than 29,000 houses were wrecked and washed away
Vancouver Island, Canada	January 26, 1700	8.7–9.2	24–30	Unknown	The length of the fault rupture was about 1,000 km with an average slip of 20 m. The tsunami killed the red cedar trees by lowering of coastal forests into the tidal zone
Seikaido-Nankaido, Japan	December 22, 1698				A large tsunami struck Seikaido-Nankaido, Japan
Bristol Channel, Great Britain	January 30, 1607		2	3,000	The tsunami caused floods that affected thirty villages in Somerset with water to a height of 1.5 m for ten days.
Keichō Nankaido, Japan	February 3, 1605	8.1	6–30	Unknown (in the order of thousands)	Earthquake of slow rupture velocity, causing little observed shaking, generated a large tsunami

(continued)

Table. 1.1 (contiuned)

Place	Date	Earthquake M	Height (m)	Death (due to ground shaking and the tsunami)	Remarks
Nueva Cadiz, Venezuela	1541			1000–1500	Possibly destroyed the population of Nueva Cadiz
Meiō Nankai, Japan	September 20, 1498	8.6	4	26,000–31,000	Tsunami was recorded in Suruga Bay and at Kamakura and ground liquefaction occurred in the Nankai area
Shōhei Nankai, Japan	August 3, 1498	8.4		660 deaths, 1700 houses destroyed	The Yunomine Hot Spring stopped and Yukiminato and Awa were completely destroyed by the tsunami
Eastern Mediterranean	August 8, 1303	About 8	9	Unknown	Severe damage and loss of life on Crete and at Alexandria
Kamakura, Japan	1293	7.1		23,000	The earthquake and tsunami destroyed Kamakura
Ninna Nankai, Japan	August 26, 887 AD				There was a strong earthquake in Osaka, Shiga, Gifu and Nagano and the tsunami flooded the coastal locality
Sendai, Japan	July 09, 869 AD	8.6		About 1000	The tsunami caused widespread flooding of the Sendai plain, with sand deposits being found up to 4 km from the coast
Hakuho, Japan	November 29, 684 AD	8.4		Unknown	The tsunami occurred off the shore of the Kii Peninsula, Nankaido, Shikoku, Kii and Awaji region

(continued)

Table. 1.1 (contiuned)

Place	Date	Earthquake M	Height (m)	Death (due to ground shaking and the tsunami)	Remarks
Alexandria, Eastern Mediterranean	July 21, 365 AD	8.5+	30+	Many thousands	The tsunami in AD 365 was so devastating that the anniversary of the disaster was commemorated annually at the end of the sixth century in Alexandria as a "day of horror"
Gulf of Naples, Italy	79 AD				Small tsunami due to the eruption of Mount Vesuvius
Helike, Greece	373 BC				The earthquake and tsunami destroyed the Greek city Helike, lying 2 km away from the sea
Malian Gulf, Greece	426 BC	Unknown	Unknown	Unknown	The Greek parts of the islands were submerged, rivers permanently displaced and towns devastated
Santorini, Greece	1600 BC	VEI 6 or 7	Eruption deposited up to 7 m		Dense-Rock Equivalent (DRE) in excess of 60 km^3 and the volume of ejecta was approximately 100 km^3
Norwegian Sea	6100 BC		Deposited sediment up to 80 km inland and 4 m above current normal tide levels		A large landslide with a volume of 3,500 km^3 of debris occurred under water causing a very large tsunami in the North Atlantic Ocean

References

McCaffery R (2008) Global frequency of magnitude 9 earthquakes. Geology 36:263–266
Murty TS, Loomis HG (1980) A new objective tsunami magnitude scale. Mar Geodesy 4:267–282

Chapter 2
1755 Lisbon Earthquake Tsunami

Abstract The November 1, 1755 Lisbon earthquake, estimated to have a magnitude of Mw 8.5 ~ 9.0, is among the worst earthquakes to have hit Europe. The estimates of human lives lost vary and extend up to 100,000. The earthquake was felt all over Europe. A major tsunami was generated that reached a height of up to 30 m. An introduction to the earthquake and an in-depth description of the tsunami are presented here.

2.1 Introduction

The fifteenth and sixteenth centuries were the golden age of Portugal when Portuguese explorers sailed all over the world, discovering Brazil, Labrador and the Cape of Good Hope. With strong naval fleets, Portuguese colonization extended to western and eastern Africa, South-East Asia, India and Brazil. A lot of wealth poured into Portugal from overseas colonies. During 1580–1581, King Phillip II of Spain conquered Portugal. However, national sovereignty was restored by the revolution of 1640. John IV ushered the silver age of Portugal during the seventeenth and eighteenth centuries when the wealth brought from Brazil made Lisbon extremely prosperous and the sought after capital of Europe. Around the middle of the eighteenth century, Lisbon was a major trading center of Europe thronged by Europeans, particularly Germans and the British.

2.2 Earthquake

The great Lisbon earthquake occurred in the morning at about 9.30 am on Saturday, 1st November, 1755 (Fig. 2.1). It was All Saints Day and most of Lisbon's Roman Catholic population was at church. The first shock shook buildings and 10 min later the second shock, which lasted some 2 min, much stronger compared to the first one, brought down the buildings. It was the second shock that did most of the damage. It destroyed palaces, tumbled church buildings, houses and shops with a deafening roar of collapsing buildings. The third shock

H. K. Gupta and V. K. Gahalaut, *Three Great Tsunamis: Lisbon (1755), Sumatra–Andaman (2004) and Japan (2011)*, SpringerBriefs in Earth Sciences,
DOI: 10.1007/978-94-007-6576-4_2, © Springer Science+Business Media Dordrecht 2013

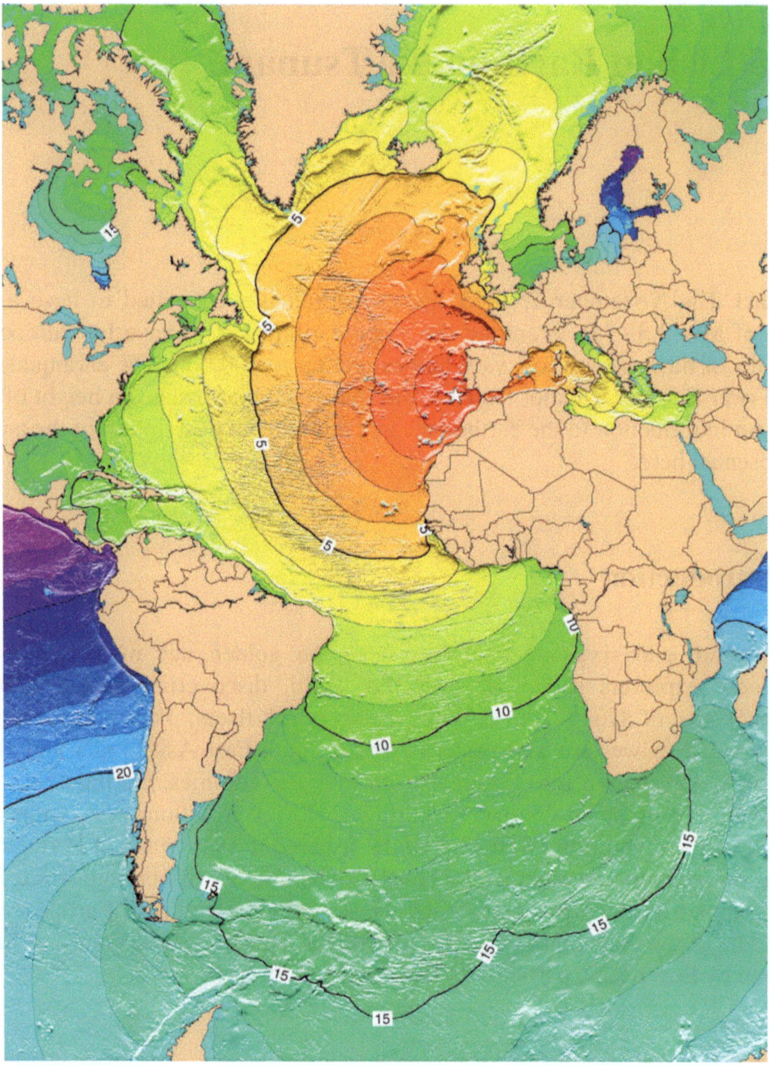

Fig. 2.1 Location of the earthquake (star) and traveltime of the tsunami. The *red* contours are for 1–4 h arrival times, *yellow* (5–6 h), *green* (7–14 h) and *blue* (15–21 h)

came soon after, completing the damage. Although there is a lot of variation in the accounts of this earthquake, it probably lasted for about 6 min. Fissures up to 5 meters wide opened up in the center of the city. Lisbon and several other major cities in Portugal were destroyed (Fig. 2.3 and 2.4). The earthquake damaged structures in Spain and Morocco. The number of human lives lost by this earthquake and the resultant tsunami continue to be argued. According to one estimate, within the first two minutes of the earthquake, some 30,000 human lives were

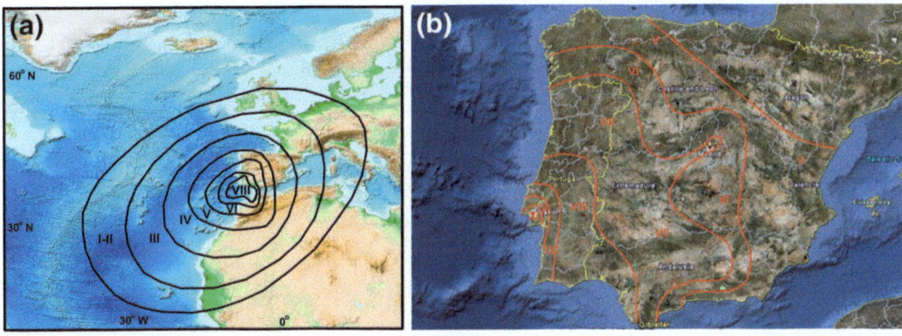

Fig. 2.2 Isoseismals of the 1755 Lisbon earthquake (Johnston 1996). The right panel shows the zoomed part of the Portuguese and Spanish regions

Fig. 2.3 Ruins of Carmo Convent

lost in Portugal. The final count is estimated to be 90,000 lives lost in Lisbon and some 10,000 lives lost in Morocco. It must be noted that at this time, the Lisbon population was 260,000. So, about one-third of the total population of Lisbon was killed. Today, the population of Lisbon is about 2,800,000. Losing some one million people in a similar tragedy today is unimaginable. The earthquake was felt all over Europe and North Africa. According to estimates of Johnston (1996), the felt area of this earthquake exceeded 14 million square km (Fig. 2.2) making it the largest documented felt area of all the world's shallow earthquakes. In Europe, ground motions from this earthquake were felt in Spain, Italy, France, Germany, Switzerland, Duchy of Luxembourg and Sweden.

Fig. 2.4 The ruins of Lisbon. Survivors lived in tents on the outskirts of the city after the earthquake, as shown in this fanciful 1755 German engraving

Fig. 2.5 Tsunami height (*vertical arrow*) and its arrival in minutes after the earthquake (*number within bracket*)

Richter (1958) has provided a very good description of the seiches set-up by the Lisbon earthquake. The earthquake set-up seiches over most of Western Europe. The farthest places were in Scandinavia and Finland. Descriptions about the seiches set-up in English harbors and ponds are documented in the 'Proceedings of the Royal Society'.

In the north dock, whose length is about 229 feet, breadth 74 feet, and at that time about 16½ feet of water, shut in by a pair of strong gates, well secured, his Majesty's ship the Gosport, of forty guns, was just let into be dock'd and well stay'd by guys and hawsers (certain large ropes, so called). On a sudden the ship ran backwards near three feet, and then forwards as much, and at the same time she alternately pitch'd with her stern and head to the depth of near three feet; and, by the liberation of the water, the gates alternately opened and shut, receding from one another near four inches…

Seiches

F.A. Forel, a Swiss scientist introduced the word "Seiches". It is typically a standing wave set on the surface of an enclosed body of water. They are mostly observed in tanks, ponds and lakes and are very often the response to passing surface waves from a distant earthquake. This figure illustrates the setting-up of seiches in an enclosed rectangular tank, where in response to the passing earthquake wave the water is made to slosh from side to side. The full and dotted lines indicate the two positions of the water surface. The illustration is for uninodal seiches. There could be multi-nodal seiches.

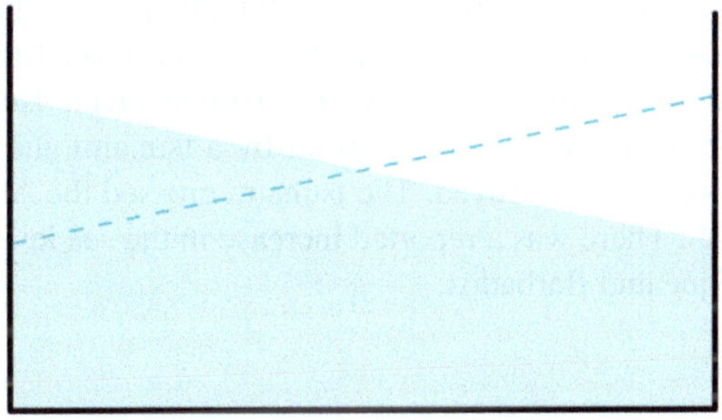

Schematic section of a seiche in a tank

2.3 Tsunami

The descriptions of the tsunamis that hit Lisbon and other parts of Portugal, Morocco and later spread all over the Atlantic Ocean are not very accurate. After going through all the reports and descriptions, one realizes the enormity of the destruction and the fear that it must have caused. It is believed that after the earthquake and the fires caused by the earthquake, many inhabitants of Lisbon looked for safety by boarding ships moored on river Tagus. After about an hour of the first shock, there was an initial recession of water in river Tagus and the upstream estuary. This exposed large portions of the river bed. However, soon after, the first of the three tsunami waves arrived. It swamped the area near Bugie Tower at the mouth of river Tagus. Moving upstream, this wave demolished Cais de Pedra at Terreiro do Poco and the nearby located custom house. In this area the wave is believed to have reached a height of 6 m. This wave also caused extensive damage to the area between Junqueria and Alcantara and the western part of the city. The boats, which were over crowded by the earthquake survivors, sank.

In all there were three waves, each dragging people and debris out to sea. The maximum run-up in height is estimated to be 20 m in the Tagus estuary.

Along the west coast: Cascais, a coastal town located some 30 km from Lisbon (Fig. 2.5), experienced an initial exposure of the sea-bed, followed by tsunami waves causing wreckage of a number of ships and loss of human lives.

In the north: The tsunami claimed many lives at Peniche, another coastal town located 80 km north of Lisbon (Fig. 2.5).

South: The water is reported to have reached a height of 4–5 m, drowning the first floors of buildings in the city of Setubal located 30 km south of Lisbon. The tsunami was particularly damaging in the southern province of Algarve, Portugal. Most of the coastal villages and towns suffered heavily. In Lagos, some 180 km south of Lisbon, the tsunami reached the top of the city walls. In the southern areas particularly, the damage inflicted by the tsunamis was much more severe than by the earthquake.

The tsunami affected Cadiz and Huelva in southwestern Spain. The tsunamis also caused wide spread damage in the western coast of Morocco extending from Tangier to Agadir (Fig. 2.5) claiming an estimated 10,000 human lives. The coasts of France, Great Britain, Ireland, Belgium and Holland also experienced tsunamis. Cornwall on the southern coast of England was hit by a three meter tsunami. Located on the west coast of Ireland, Galway was also hit by a tsunami and the 'Spanish Arc' section of the city wall was destroyed. The tsunami crossed the Atlantic and reached Antilles after 6 hours. There was a reported increase in the sea level by about a meter at Antigua, Martinique and Barbados.

2.4 Fire

In addition to earthquakes and tsunamis, Lisbon was also destroyed by fires, which soon set in after the earthquakes (Fig. 2.6). These fires were started mostly by cooking fires and candles. In the densely populated areas these fires spread very quickly. As the people were running away from their homes after the earthquake, no one had time to attend to these fires and they spread very fast. Moreover, the narrow streets were filled with debris from the falling of buildings making the access to fire sites difficult. Within a short time the fires reached public squares, where inhabitants had gathered with their belongings. Buildings, which had not suffered too much damage by the earthquake, were destroyed by the fires. The Opera House and the Royal Palace burnt down. The Patriarchal Church, which was not severely damaged by the earthquake and was continuing to provide

Fig. 2.6 This 1755 copper engraving shows the ruins of Lisbon in flames and a tsunami overwhelming the ships in the harbor

religious services, had to be abandoned as the fires approached. Finally, it was totally gutted. The fire continued for almost 5 days. There is no separate estimate available about the total lives lost due to the fire.

2.5 Societal Response

The earthquake occurred on an important Catholic religious holiday, All Saints Day, in the morning when a large number of religious people were inside churches. Many of them were killed with the collapse of most of the important churches in Lisbon. This rattled the faith of staunch Roman Catholics. Was it the wrath of God? Preachers urged devout Catholics to pray and not to commit any sin. Were the earthquake of November 1, 1755 and its continued aftershocks due to the anger of God? A leading Jesuit, Malagrida, was a very successful preacher and he preached against the notion that earthquakes were due to natural causes in his sermons. His message to people was for humility and repentance to God during the nine month period of aftershocks (Livermore 1976). He insisted that this was not the time for rebuilding and reconstruction as God was still angry. The eighteenth century was characterized by the end of the religious wars of the seventeenth century. There was an enormous desire all over Europe that everything should develop peacefully. Trade, commerce, philosophy and natural sciences contributed to a stable world. Isaac Newton (1642–1727) had discovered gravity and laws to predict planetary movement. Gottfried Wilhelm Leibniz (1646–1718), along with Newton, had invented differential calculus and was trying to expand the notion of optimization from mathematical functions and physics into metaphysics. He enquired philosophically "Why is there something rather than nothing?" and "Why is it, as it is?". Through his essay, "*Theodic'ee on the Goodness of God, the Freedom of Men, and the origin of Evil*" he concluded "That this must the best of all possible worlds". This was the time in the eighteenth century when almost all of Europe believed in Leibniz's philosophy. The British poet Alexander Pope (1688–1744) expressed it concisely by saying "What is, is good". The Lisbon earthquake shattered the notion of, "Best of the all possible worlds".

The intelligentsia of the European Age of Enlightenment was strongly influenced by the Lisbon earthquake. Voltaire (1694–1778) and Kant (1724–1804) were struggling to show earthquakes as a natural phenomenon rather than an act of God. The news of the earthquake reached Voltaire in Geneva, by stage coach, 15 days after the earthquake and he attacked Leibniz's theorem that this was the best world. In 1756 Immanuel Kant published the first scientific description of the Lisbon earthquake, "The Earthquake which shook at the end of 1755th year large parts of the Earth".

2.6 Recovery and Reconstruction

It is interesting to note that King Joseph I, the entire royal family and Prime Minister Sebastiao de Melo (the Marquis of Pombal) survived the catastrophe. King Joseph I became claustrophobic after the earthquake and this resulted in holding court in the open in a complex of tents on the hills of Ajuda. Prime Minister Pombal was a very clever man. When asked, what needs to be done, he is reported to have responded, "Bury the dead and heal the living" (Kendrick 1957). This was a very well organized response, providing relief and rehabilitation to the people who had suffered. The first most important thing was to douse the fires. Firefighters were sent to all the places where fires were destroying whatever was left from the havoc created by the earthquake and the resultant tsunami. The next important issue was to dispose of tens of thousands of corpses spread all over before the spread of disease. To the dislike of religious individuals and contrary to the custom, a large number of dead bodies were loaded into barges and buried at sea. Law and order was maintained through deployment of the army. Large-scale looting was prevented and the perpetrators were sent to gallows erected at central places in the city. It is reported that more than 30 culprits were hanged (Gunn 2008). The army also prevented able bodied Portuguese to leave Lisbon, enlisting them into recovery and reconstruction work.

After overcoming the immediate problems of the fires, disposing the dead bodies and putting the royal administrative machinery back into shape, the next most important thing that needed attention was the rebuilding of Lisbon. It is noteworthy that just one month after the earthquake, on 4th December 1755, Manuel da Maia, the chief engineer presented to King Joseph I three possible options: 1. Rebuild the old city by repairing the damaged buildings using recycled material,

Fig. 2.7 Parca Do Comercio, centerpiece for Pombal's new grid design

2. Widening of certain streets and 3. Razing the entire Baixa quarter and laying down new streets without restraints (Shrady 2008). The King and Prime Minister Pombal opted for the third alternative. All the debris was removed within one year and new construction with wide roads, big squares and large rectangular avenues was undertaken. They found a way to construct earthquake proof houses by making wooden models of buildings and marching troops around them to simulate an earthquake and test the performance of the wooden models. This newly developed down town, Pombaline Downtown, is today one of the tourist attractions of Lisbon (Fig. 2.7).

2.7 Birth of Seismology

Following the Lisbon earthquake, the disaster management undertaken had systematic steps that are still used for collecting quantitative information about earthquakes (Shardy 2008; Fuchs 2008). Questionnaires were sent to convents, priests and officials. The questions were:

- At what time did the earthquake begin and how long did it last?
- How many shocks were felt?
- Did you perceive the shock to be greater from one direction than another? For example, from the north to the south? Did buildings seem to fall more to one side than the other?
- How many people died and was any one of them distinguished?

- Did the sea rise or fall first and how many hands did it rise above the normal?
- If fire broke out, how long did it last and what damage was caused?

Even today, similar questions are asked after a major or damaging earthquake, to prepare earthquake intensity distribution maps.

Torre do Tombo, the national historical archive at Lisbon has preserved answers to these and similar questions. In the absence of the questionnaire designed by Marquis of Pombel, it would not have been possible for researchers today to rework the dynamics of the Lisbon earthquake using modern day tools of analysis and interpretation. Hence, Marquis of Pombel is credited to be the forerunner of seismological developments that followed the Lisbon earthquake.

2.8 Recent Investigations

The magnitude of the 1755 Lisbon earthquake has been debated and examined in the past few decades. The Royal Academy of History (RAH) of Spain had collected very detailed information on the effects of the Lisbon earthquake in Spain within one year of the occurrence of the earthquake. Using these data, Solares et al. (1979) derived MSK intensity values for over 1000 locations to draw an isoseismal map for Spain. Figure. 2.2 shows the isoseismals maps for Portugal and Spain adopted from Solares et al. (1979). Johnston (1996) used the available data as well as observations from the ships records that were in the Atlantic Ocean, to reconstruct regional isoseismals spreading over Europe. Johnston also estimated the magnitude of the Lisbon earthquake to be ~8¾. The mechanism and the source of the Lisbon earthquake have been investigated by several authors in the last decade (for example Gutscher 2004 and Baptista et al. 2003). We quote Karl Fuchs (2008) "The European plate is colliding with the approaching African plate. A thrust movement on a plane of about 16,000 km^2 broke with a relative movement of 12 m. This elevated the sea floor and generated the tsunami which reached Lisbon with a height of about 7 m in the trumpet like mouth of Tejo".

References

Baptista MA, Miranda JM, Chierici F, Zitellini N (2003) New study of the 1755 earthquake source based on multi-channel survey data and tsunami modelling. Nat Hazards Earth Sci (special issue) 3:233–240

Fuchs K (2008) The great earthquakes of Lisbon 1755 and Aceh 2004 shook the world-seismologists' societal responsibility. In: Ribeiro A, Mendis-Victor L, Sousa Olivera C, Azevedo J (eds) The Lisbon earthquake. Revisited. Springer Netherlands

Gunn AM (2008) Encyclopedia of Disasters: environmental catastrophes and human tragedies. Greenwood Press, Westport CT, p 735

Gutscher MA (2004) What caused the Great Lisbon earthquake? Science 305:1247–1248

Johnston AC (1996) Seismic moment assessment of earthquakes in stable continental regions-III, New Madrid 1811–1812, Charleston 1886 and Lisbon 1755. Geophys J Int 126:314–344

Kendrick TD (1957) The Lisbon Earthquake. J.B. Lippincott Co., Philadelphia

Livermore HV (1976) A new history of Portugal. Cambridge University Press, Cambridge

Richter CF (1958) Elementary seismology. W.H. Freeman & Co., San Francisco p 768

Shrady N (2008) The last day: wrath, ruin and reason in The Great Lisbon Earthquake of 1755, Penguin, p 256

Solares JM, Martinez A, Lopez A, Mezcua J (1979) Isoseismal map of the 1755 Lisbon earthquake obtained from Spanish data. Tectonophysics 53:301–313

Chapter 3
2004 Sumatra–Andaman Earthquake and Tsunami

Abstract The December 26, 2004 Sumatra earthquake of Mw 9.1 caused wide spread damage in South and South-East Asian countries. The resultant tsunami claimed an estimated 230,000 human lives, the largest ever loss of human lives in a tsunami. Extensive damage and loss of human lives occurred in Indonesia, Thailand, Sri Lanka, India and many other countries. This chapter includes an introduction to the earthquake and an in-depth description of the tsunami.

3.1 Introduction

In the previous chapter we discussed the 1755 Lisbon tsunami and the birth of the science of earthquakes (seismology). A lot of new discoveries were made in the next 200 years, the most significant among them being the development of traveltime tables of waves generated by earthquakes and the structure of the Earth deduced from the observation of earthquake waves traveling through and around the Earth. A major development was the setting up of the World Wide Standard Seismograph Network during 1963–1964, where over 100 seismic stations equipped with three-component short-period Benioff seismographs and three-component long-period Press Ewing seismographs were set up, providing for the first time a global coverage of earthquake recordings using similar instruments. This also led to the development of the plate tectonics hypothesis during the late 1960s and early 1970s. Seismological networks further expanded tremendously during the last three decades of the twentieth century. Seismological arrays were installed in Europe, Japan and America. Crustal deformation studies using GPS measurements provided impetus to our knowledge about plate motion and crustal deformation due to earthquake occurrence processes. It appeared as if the seismologists were just getting ready to study great earthquakes that were to occur over the next decadal years after a general lull of mega earthquakes (Mw 9) since the great 1964 Alaska earthquake of Mw 9.4. This lull was broken by the mega Andaman–Sumatra earthquake on 26 December 2004, on Boxing Day, the day after Christmas. Various estimates of its magnitude range from 9.0 to 9.3, however, a magnitude estimate of 9.1 is considered as the most robust. The region of its occurrence was not considered to produce such a large earthquake and hence the tsunami caused by this earthquake devastated a large area around the source region. Thus

H. K. Gupta and V. K. Gahalaut, *Three Great Tsunamis: Lisbon (1755), Sumatra–Andaman (2004) and Japan (2011)*, SpringerBriefs in Earth Sciences,
DOI: 10.1007/978-94-007-6576-4_3, © Springer Science+Business Media Dordrecht 2013

this earthquake and the tsunami took people by surprise. In the known history of the coastal region around the Bay of Bengal and even around the Indian Ocean, the tsunami caused by this earthquake was the most devastating, killing about 230 thousand people. Nobody expected it to have that far a reach, killing people more than 5000 km away in Somalia. Maximum damage was in Sumatra, Indonesia, India's Andaman and Nicobar Islands and east coastal region and Sri Lanka.

3.2 Tectonics and Earthquake History in the Sunda Arc

The Sumatra–Andaman subduction zone marks the eastern boundary of the Indian plate where it subducts under the Sunda plate (Fig. 3.1). The 26 December 2004 giant Sumatra–Andaman earthquake occurred in this subduction

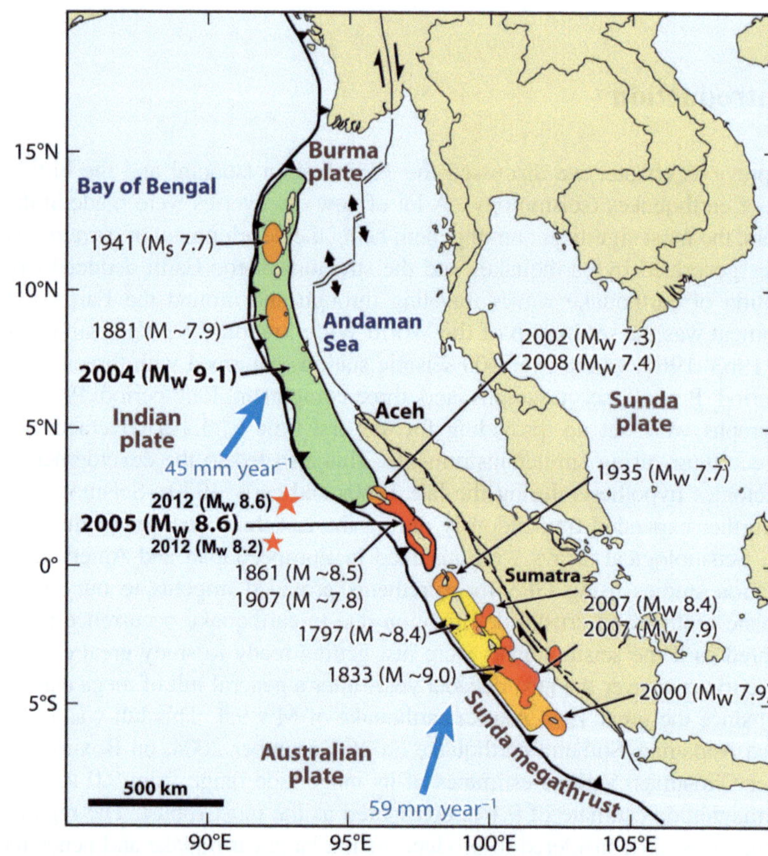

Fig. 3.1 Earthquake ruptures in the Sumatra–Andaman subduction zone. The inferred magnitudes for the historical earthquakes are denoted by M. Ms is the estimated surface wave magnitude whereas Mw is the moment of magnitude estimated for recent earthquakes. The blue arrows indicate the rate and orientation of plate motions relative to the Sunda plate (modified from Shearer and Burgmann 2010)

zone. No great earthquake (M $\geq$ 8) has been reported from the Andaman–Nicobar and northern Sumatra regions, though major events in 1847 (M 7.5), 1868 (M 7.6), 1881 (M 7.9) and 1941 (M 7.7) occurred in the regions. Great earthquakes in 1797, 1861, 1833, 2005 and 2007 have been reported from the subduction zone near and southeast of Sumatra. Recently, on 11 April 2012, two great earthquakes of unprecedented magnitudes of 8.6 and 8.2 occurred about 100–200 km west of the subduction zone in the Sumatra region (Fig. 3.1). As these earthquakes did not occur in the subduction zone, they are considered as intraplate earthquakes and are probably the largest magnitude intraplate strike-slip earthquakes recorded globally.

3.3 The 2004 Sumatra–Andaman Earthquake

The 2004 Sumatra earthquake nucleated off the western coast of northern Sumatra and propagated north-northwest along the subduction zone right up to the North Andaman Island. Thus, the rupture length of the earthquake was about 1400 km (Fig. 3.1). This was the longest rupture ever reported for any earthquake (Fig. 3.2). Other than the extraordinary large rupture length, the earthquake had a few more distinctive features. Generally, the rupture speed during an earthquake is about 2.5 km/s, which is almost equal to the shear-wave velocity. The southern part of this earthquake rupture exhibited normal speed with a magnitude of slip reaching 20 m. However, the northern part of the rupture, under the Andaman Islands, exhibited a slower rupture speed (Fig. 3.3). Because of the slow rupture in the northern part, the seismological data do not constrain the slip on the rupture under the Andaman and Nicobar Islands reasonably well, as most of the slip in this part occurred at a time scale beyond the seismic band (Lay et al. 2005, Ammon et al. 2005) and therefore

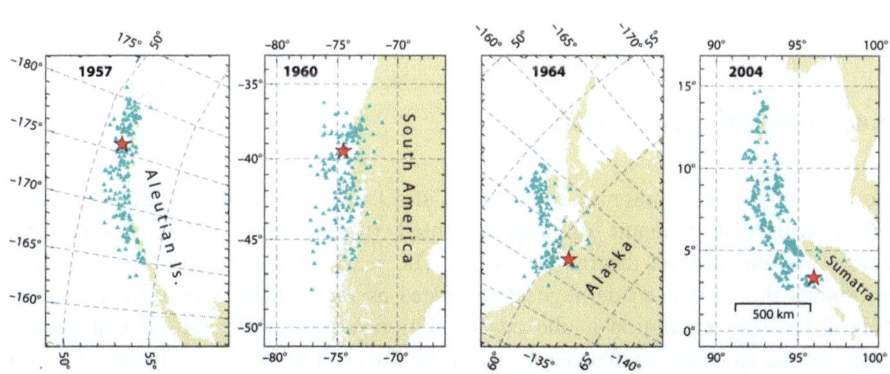

Fig. 3.2 Rupture lengths of the 1957, 1960, 1964 and 2004 earthquakes (Mw $\geq$ 9), which are based on the epicentral distribution of the aftershocks that occurred within one month after the mainshock. Red stars show the mainshock epicenters. The scale is the same on all the maps. Note the longest rupture length is for the 2004 Sumatra–Andaman earthquake. Figure after Ishii et al. (2005)

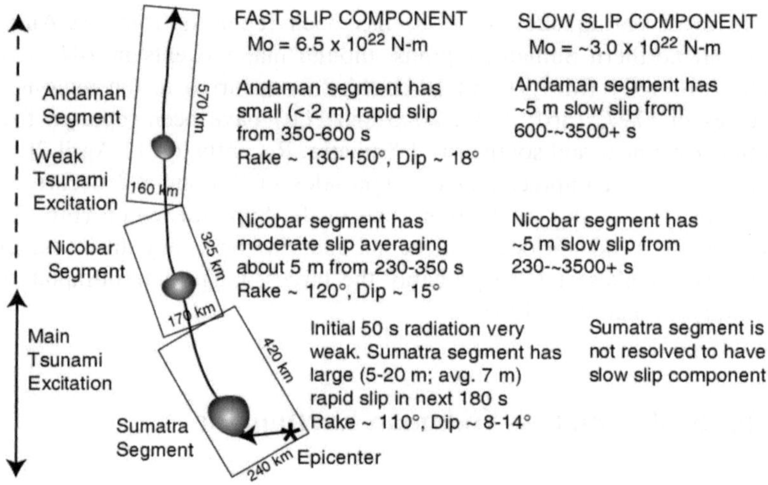

Fig. 3.3 Three stages of rupture propagation of the 2004 Sumatra–Andaman earthquake (after Lay et al. 2005)

uncertainty in the estimate of the magnitude prevails. This earthquake was the best monitored earthquake until that time, as almost all the instrumentation were in place to record such an imposing event. The GPS data provided additional information about the rupture length and slip (Gahalaut et al. 2006). It was only after the occurrence of this earthquake that seismologists realized the problem in estimating earthquake magnitude and the energy released during exceptionally large earthquakes (Mw > 8.5). For this earthquake, the estimates based on energy released by surface waves of up to a 500 s period, led to an Mw estimation of 9.0. However, due to the high-quality seismic data available for this earthquake, seismologists could estimate the energy released by the waves of periods >500 s, which caused the revision of the magnitude by 0.1–0.3 units (Fig. 3.4).

3.4 Tsunami Generation

Because of its large magnitude and huge slip (>20 m) on the rupture, this earthquake caused a tsunami of unprecedented magnitude, which devastated the coastal regions around the Indian Ocean. The reach of the tsunami waves was so enormous that it caused damage at distances as far as 6000 km from the source (Fig. 3.5). The population in the coastal region around the Indian Ocean has not experienced tsunamis as frequently as in the Pacific region and hence the tsunami took people by surprise. This resulted in a huge loss of human lives and damage to coastal establishments. There is some debate on the source of the tsunami. The tsunami generation process depends on rupture speed. Typical rupture speed is the shear-wave speed in rocks. In case the rupture speed is slow, then the process of

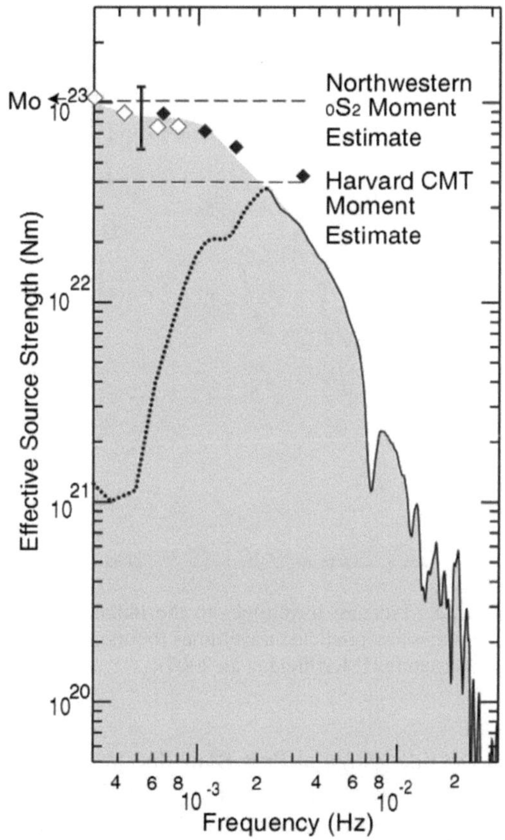

Fig. 3.4 Estimation of the seismic moment of the 2004 Sumatra–Andaman earthquake. The conventional Harvard CMT moment estimate is 4×10^{22} Nm, whereas the estimate obtained from lower frequency (longer period) surface waves is about 10^{23} Nm, which corresponds to a magnitude of 9.3 (after Stein and Okal 2005)

coseismic elevation changes of the sea-bed would be slow and hence the water will not be displaced vertically. In the case of the 2004 Sumatra–Andaman earthquake, it appears that the northern part of the rupture did not contribute to the tsunami generation. Lay et al. (2005) analyzed the seismological data and rupture process of the earthquake. They suggested that about one-third of the total seismic moment was released due to a slow slip. The Sumatra rupture segment (about 420 km) did not exhibit a slow slip but in the Nicobar region (rupture segment of about 325 km), about one half of the total slip occurred through a slow slip. In the Andaman segment (about a 570 km long rupture segment) the slip was predominantly slow. Shearer and Burgmann (2010) provided a nice comprehension of possible tsunami models of the 2004 Sumatra–Andaman earthquake. All these models point towards an at least 1400 km long rupture that extended from the point where the earthquake initiated to the North Andaman Island. However, there is some disagreement in the estimation of the tsunami source, primarily because of lack of constraints on the extent of the slow slip in the northern part. This could easily be settled if there were near-source tide gauge instruments in the northern part of the rupture. Although there was a tide gauge at Port Blair, unfortunately, there was a timing problem

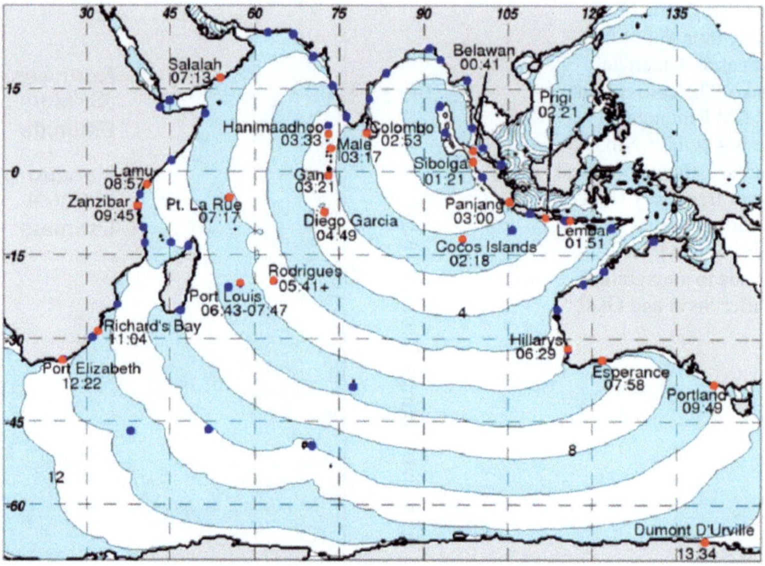

Fig. 3.5 Tsunami traveltimes to the Indian Ocean tide gauge stations, in hours:minutes. The contours show predicted traveltimes (hours). The red and blue dots show the locations of the tide gauge stations (Merrifield et al. 2005)

at the tide gauge at Port Blair due to which the slow slip component in the region could not be resolved properly. Singh et al. (2006) attempted to rectify the timing errors in the tide gauge record at Port Blair and their preferred model consists in a mixed mode of slip in which about one half of the total slip occurred seismically in less than 5 min and the rest in the next 30 min. The existing tsunami modeling studies are not in complete agreement regarding the length of the best-fitting source region and the rupture velocity (Shearer and Burgmann 2010).

3.5 Description of the Tsunami

The 2004 tsunami affected a large region around the Indian subcontinent. The reach of the tsunami (beyond 5000 km) and its wave height at such large distances were unprecedented. Even more so because this region was not considered to be prone to a tsunamigenic earthquake, though a few cases of a tsunami, e.g., due to the 1945 Makran earthquake, 1833 Krakatoa volcanic eruption, etc., had been reported. Almost all the damage and loss of life occurred due to the tsunami caused by the earthquake, rather than shaking, despite the large magnitude of the earthquake. Even in the source zone in the Sumatra–Andaman Island region, the damage was mostly due to the tsunami. It was one of the deadliest natural disasters in recorded history. Indonesia was the hardest-hit country, followed by

Sri Lanka, India and Thailand. Because of the 1,400 km long rupture with an almost north–south orientation, the greatest strength of the tsunami waves was in an east–west direction. Because of the distances involved, the tsunami took anywhere from fifteen minutes to seven hours (for Somalia) to reach the various coastlines. The northern regions of the Indonesian island of Sumatra and the Andaman and Nicobar Islands of India were hit very quickly, while Sri Lanka and the east coast of India were hit roughly 90 min to two hours later (Fig. 3.5). Thailand was also struck about two hours later despite being closer to the epicenter, because the tsunami traveled slowly in the shallow Andaman Sea off its western coast. The tsunami was noticed as far away as Struisbaai in South Africa, some 8,500 km away, where a 1.5 m high wave surged on shore about 16 h after the earthquake. It took a relatively long time to reach this spot at the southernmost point of Africa, probably because of the broad continental shelf off South Africa and because the tsunami would have followed the South African coast from east to west. The tsunami also reached Antarctica, where tidal gauges recorded oscillations of up to a meter, with disturbances lasting a couple of days. Some of the tsunami's energy escaped into the Pacific Ocean, where it produced small but measurable tsunamis along the western coasts of North and South America, typically around 20–40 cm.

It is estimated that about 230 thousand people in the coastal region died due to the tsunami (Table 3.1). This makes it the worst tsunami in history.

3.5.1 Tsunami in the Open Ocean

The US-French satellites, TOPEX/Poseidon and Jason-1, passed over the Bay of Bengal two hours after the earthquake and captured the height of the propagating tsunami. It was 60 cm high. By 3 h 15 min after the earthquake, the height

Table 3.1 Countrywide death tolls due to the 2004 tsunami

Country where deaths occurred	Confirmed	Estimated
Indonesia	126,915	167,799
Sri Lanka	30,196	35,322
India	10,610	18,045
Thailand	4,812	8,212
Somalia	298	298
Myanmar	61	400–600
Maldives	82	108
Malaysia	68	75
Tanzania	10	13
Seychelles	3	3
Bangladesh	2	2
South Africa	2	2
Yemen	2	2
Kenya	1	1

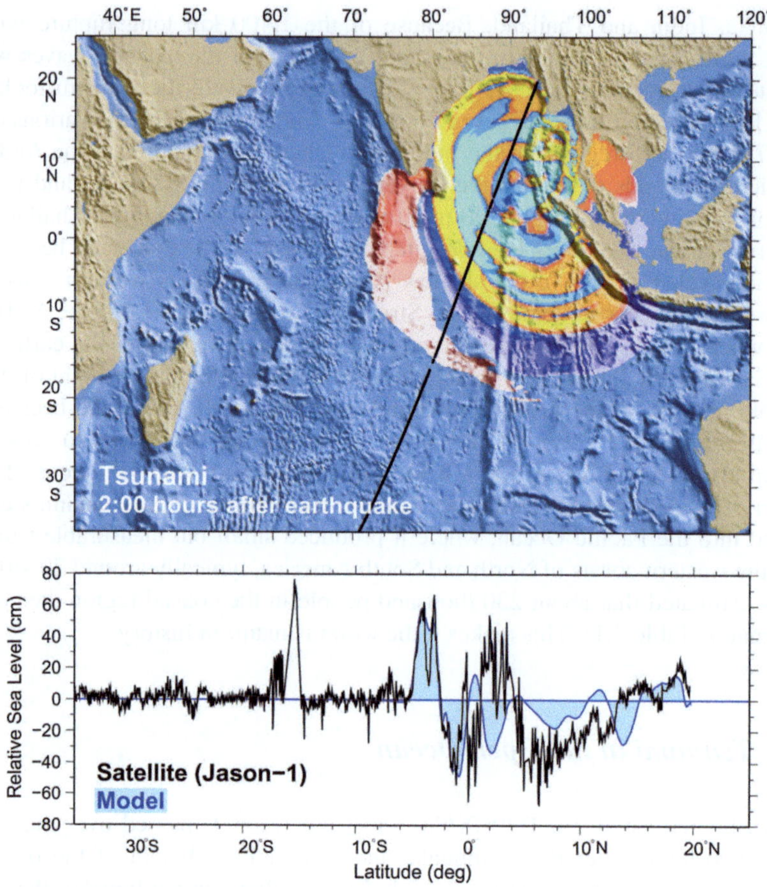

Fig. 3.6 The tsunami wave height as measured by satellites two hours after the earthquake. Lower panel shows the tsunami wave height along the satellite track, shown by black line in the upper panel (after NOAA)

dropped to around 40 cm. By 8 h 50 min after the earthquake, the wave spread over most of the Indian Ocean and was quite small in most areas, with height as small as 5–10 cm (Fig. 3.6).

This tsunami was observed at several tide gauge stations located in the Indian Ocean (Figs. 3.7 and 3.8).

3.5.2 Tsunami in the Indian Coastal Region and on the Islands

According to official estimates in India, 10,136 people were killed and hundreds of thousands were rendered homeless. The most affected regions in India were the Andaman and Nicobar Islands and the eastern coast.

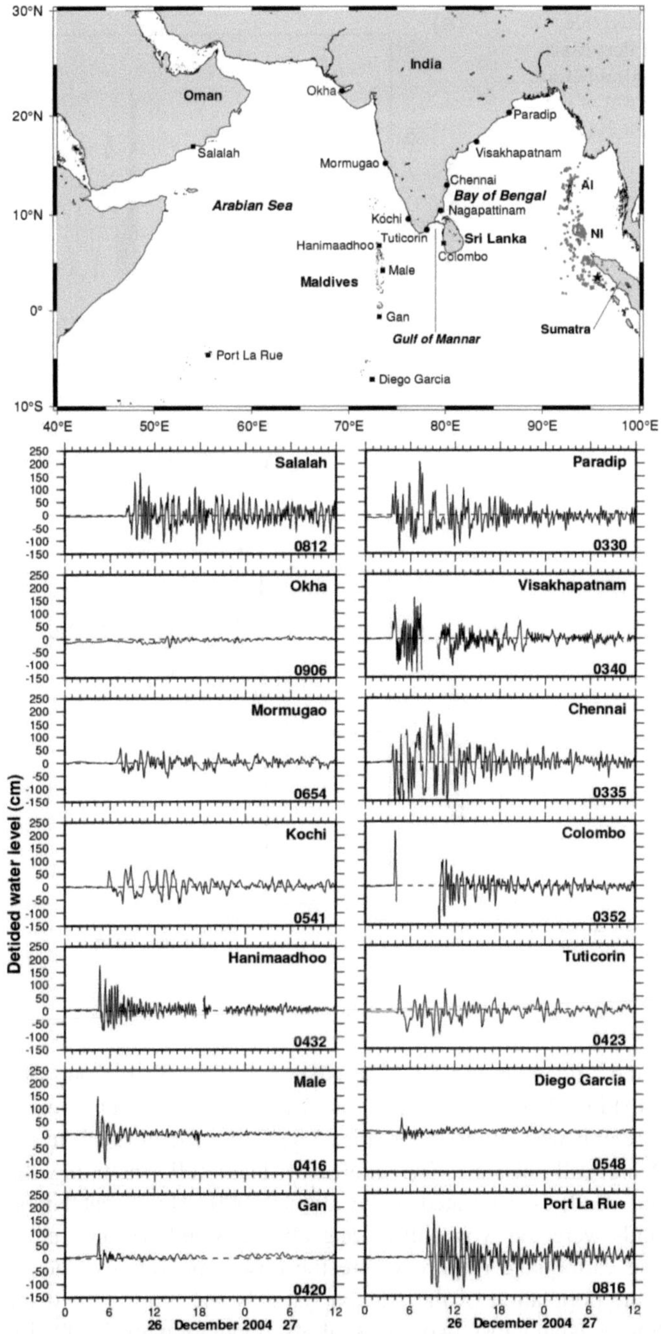

Fig. 3.7 Detided records at various tide gauge stations showing the tsunami (Nagrajan et al. 2006). The epicenter of the 26 December earthquake (*asterisk*) and the locations of the aftershocks (*grey circles*) till 10 February 2005 are also shown on the top panel. *AI* Andaman Islands; *NI* Nicobar Islands. The time at the right bottom in each box is in UTC

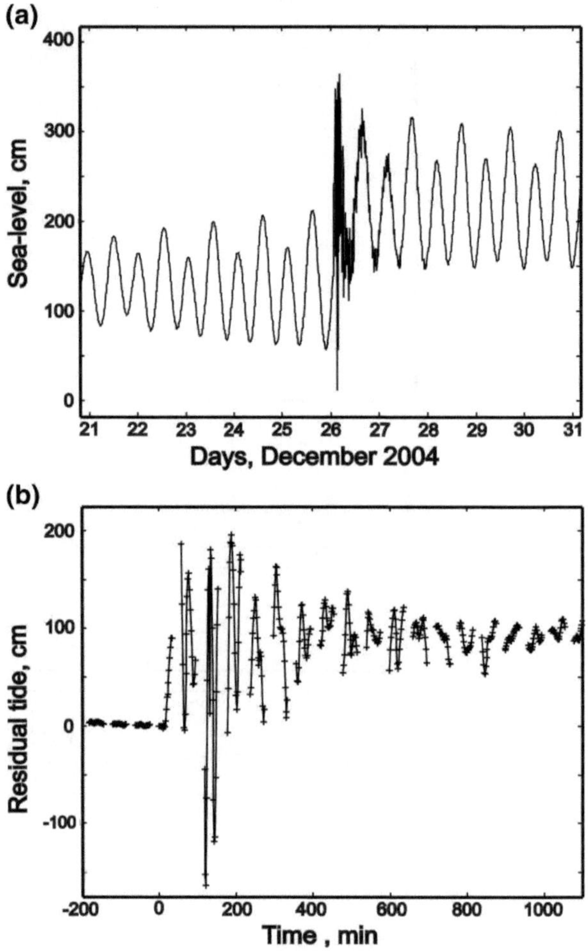

Fig. 3.8 Corrected tide gauge record at Port Blair (Andaman Island) **a** before and **b** after removal of the tides (Singh et al. 2006)

Amongst the Andaman and Nicobar Islands, the Nicobar group of islands was most affected (Figs. 3.10, 3.11, 3.12, 3.13, 3.14, 3.15, 3.16). The height of the tsunami in this region was about 15 m (Fig. 3.9). The official death toll is 1,310 and about 5,600 are still missing. The unofficial death toll (including those missing and presumed dead) is estimated to be about 7,000. The Great Nicobar and Car Nicobar Islands were the worst hit among all the islands as this region is located just above the earthquake rupture where the coseismic slip was at its maximum. One-fifth of the population of the Nicobar Islands was said to be dead, injured or missing. Chowra Island lost two-thirds of its population of 1,500. Entire islands were washed away and the island of Trinket split into two islands (NRSC). Among the casualties in Car Nicobar, 111 Indian Air Force personnel and their family members were washed away when the wave hit the air base. Unfortunately,

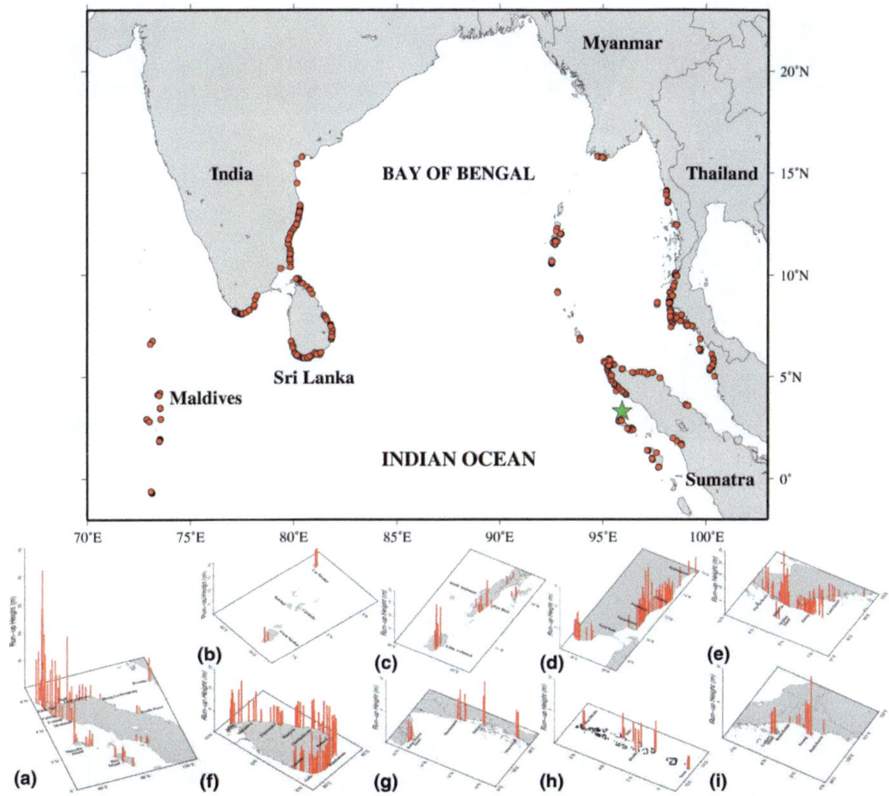

Fig. 3.9 Locations and run-up wave heights of the measured points by the KSCOE survey team and the International Tsunami Survey Teams (a-i). Star indicates the 2004 Sumatra–Andaman earthquake. **a** Indonesia; **b** Nicobar Islands, India; **c** Andaman, India; **d** east coast of India; **e** Thailand; **f** Sri Lanka; **g** Myanmar; *h* Maldives; *i* Malaysia (after Choi et al. 2006)

the topography of all these islands is quite flat and hence it did not provide any resistance to the tsunami. In many cases, tsunami waves swept through entire islands.

The eastern coast of the Indian mainland (particularly in the Tamilnadu and Andhra Pradesh states) was severely affected (Fig. 3.17). It took about 2 h for tsunami waves to travel to the Indian east coast. In Tamil Nadu, the Nagapattinam-Cuddalore shelf was the worst affected by the tsunami where more than 7800 people died. Tsunami heights in this part were of the order of 2–5 m, with an inundation of 150–800 m into the interior coast, thus causing a huge loss of human life and property. The main reason for the great loss of lives and property is due to its relative proximity to the origin of the event, apart from the concave nature of the shelf with a gentle gradient. In Andhra Pradesh, more than 100 people died. The tsunami encroached up to 500 m to 2 km due to its flat region. The tide gauge at Vishakhapatnam showed a tsunami height of 1.4 m, though eyewitnesses reported a height of up to 5 m at some other locations.

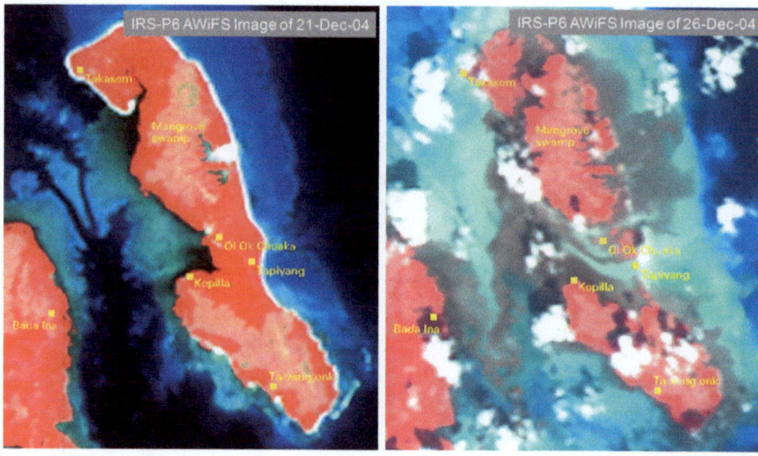

Fig. 3.10 IRS P6 AWiFS images of Trinkat Island (Nicobar) before (21 December 2004) and after the tsunami (26 December 2004) (*Image courtesy* National Remote Sensing Centre)

Fig. 3.11 The air base of the Indian Army at Car Nicobar Island was worst hit by the tsunami

3.5.3 Tsunami in Indonesia and Thailand

The northern and western coast of Sumatra, Indonesia and smaller islands west of Sumatra, were seriously affected by the earthquake and the tsunami. There were more than 31,000 casualties and most of the damage took place within

Fig. 3.12 The Ashton creek bridge in North Andaman was offset by more than a foot by extensive ground shaking

Fig. 3.13 A log that was transported by the tsunami waves got trapped in the trees at Great Nicobar

Fig. 3.14 A photo combo of the light house at Indira point at Great Nicobar Island before and after the earthquake. Coseismic subsidence of more than 1 m and extreme inundation due to the tsunami can be seen here

the province of Aceh (Figs. 3.18, 3.19, 3.20). According to the country's National Disaster Relief Coordination Agency, 126,915 people died and 37,063 are missing. This region was the closest to the earthquake epicenter and rupture and thus people had very little lead time. Casualties in the Nias and Simuelue Islands were not very high despite strong ground shaking and a high tsunami of more than 10 m. This is mainly because the natives were aware of the hazards of a tsunami.

Fig. 3.15 Mud-volcanoes erupted in the Middle Andaman Islands and one of them caught fire

Fig. 3.16 There were large cracks in the mud-volcano region and in one case, the trunk of a tree got split

Fig. 3.17 Devastation after the tsunami at Chennai, on the east coast of India

The Thai government reported 4,812 confirmed deaths and 4,499 missing after the country was hit by the tsunami. The popular tourist resort of Phuket was badly hit with 3,950 confirmed deaths. Quite a lot of these casualties are due to the fact that it is a very popular tourist destination and the tsunami struck this region at about 10–11 am on a Sunday morning when most of the people were on the beaches.

3.5.4 Tsunami in Sri Lanka

The entire eastern coast of Sri Lanka was severely affected by the tsunami. Sri Lankan authorities reported 30,196 confirmed deaths. About 1,200 people died at Batticaloa in the east. At Trmcomalee in the northeast, where the tsunami encroached more than 2 km inland, 800 people were reported dead. In the neighboring Amparai district alone, more than 5,000 people died. Tragically, a holiday train, the "Queen of the Sea", was struck by the tsunami near the village of Telwatta as it traveled between Colombo and Galle carrying at least 1,700 passengers, killing most of them (Fig. 3.21). The tsunami height at several places was about 10 m.

3.5.5 Tsunami in Somalia

The reach of the 2004 tsunami was really extensive. Somalia, which is 5000 km beyond the earthquake epicenter, was badly hit by the tsunami. The confirmed

Fig. 3.18 Satellite image of Banda Aceh, before and after the earthquake

death toll is 298. Most of the damage was in the coastal region of Puntland, particularly the area between Hafun in the Bari region and Garacad in the Mudug region. The narrow and low-lying peninsula of Hafun, 1,150 km northeast of Mogadishu, was particularly devastated.

3.6 Lessons Learnt

1. One of the most important aspects of the 2004 Sumatra–Andaman earthquake is the realization of the seismogenic and tsunamigenic potential of low-strain accumulating regions. Simple relations between rate, slab age, or structure of a subduction zone and the maximum size of events can be misleading.

Fig. 3.19 A barge lying 3 km inside Sumatra after the tsunami

Fig. 3.20 Total destruction at Banda Aceh

Fig. 3.21 The holiday train "Queen of the sea" was struck by the tsunami killing almost all the passengers

Thus, all areas of active subduction should be considered at risk of being hit by great earthquakes (Burgmann and Shearer 2010).

2. The occurrence of this and other recent great earthquakes requires improved methods to quickly estimate the earthquake size and the tsunami potential worldwide to provide better advance warning of the hazards from future megathrust earthquakes.
3. The tsunami caused by the 2004 earthquake warns us of a possible longer reach, high run-up height and heavy destruction by tsunamis.

References

Ammon CJ, Ji C, Thio HK, Robinson D, Ni SD et al (2005) Rupture process of the 2004 Sumatra–Andaman earthquake. Science 308:1133–1139

Choi BH, Hong SJ, Pelinovsky E (2006) Distribution of runup heights of the December 26, 2004 tsunami in the Indian Ocean. Geophys Res Lett 33:L13601. doi:10.1029/2006GL025867

Gahalaut VK, Nagarajan B, Catherine JK, Kumar S (2006) Constraints on 2004 Sumatra–Andaman earthquake rupture from GPS measurements in Andaman–Nicobar islands. Earth Planet Sci Lett 242:365–374

Ishii M, Shearer PM, Houston H, Vidale JE (2005) Extent, duration and speed of the 2004 Sumatra–Andaman earthquake imaged by the Hi-Net array. Nature 435:933–936

Lay T, Kanamori H, Ammon CJ, Nettles M, Ward SN et al (2005) The great Sumatra–Andaman earthquake of December 26, 2004. Science 308:1127–1133

Merrifield MA et al (2005) Tide gauge observations of the Indian Ocean tsunami, December 26, 2004. Geophys Res Lett 32:L09603. doi:10.1029/2005GL022610

Nagarajan B, Suresh I, Sundar D, Sharma R, Lal AK, Neetu S, Shenoi SSC, Shetye SR, Shankar D (2006) The great tsunami of 26 December 2004: a description based on tide-gauge data from the Indian subcontinent and surrounding areas. Earth Planets Space 58:211–215

Shearer P, Burgmann R (2010) Lessons learnt from the 2004 Sumatra Andaman megathrust rupture. Annu Rev Earth Planet Sci 38:103–131

Singh SK, Ortiz M, Gupta HK, Ramadass DGA (2006) Slow slip below Port Blair, Andaman, during the great Sumatra-Andaman earthquake of 26 December 2004. Geophys Res Lett 33:L03313. doi:10.1029/2005GL025025

Stein S, Okal EA (2005) Speed and size of the Sumatra earthquake. Nature 434:581–582

Chapter 4
2011 Tohoku-Oki Earthquake and Tsunami

Abstract The March 11, 2011 Tohoku earthquake of Mw 9.0 was a surprise to seismologists in Japan and globally. This earthquake and the resultant tsunami claimed about 20,000 human lives and caused wide spread damage to structures. The tsunami also caused a number of nuclear accidents. This earthquake gave rise to a global debate on the anticipated maximum size of earthquakes and the safety of nuclear power plants globally. This chapter includes a discussion on the Mw 9.0 earthquake, an in-depth analysis of the generation and propagation of the tsunami and a brief description of the damage to nuclear power plants and the future plans for protection.

4.1 Introduction

Japan is known for earthquakes and tsunamis and that is why the Japanese language word "tsunami" is so popular and used all over the world. Japan is located near the subduction zone formed by the Pacific, North America, Philippines and Eurasian plates. The Pacific plate moves approximately westwards with respect to the North America plate at a rate of ~8 cm/yr and subducts beneath Japan at the Japan Trench (Fig. 4.1). In the north of Tokyo, earthquakes are caused by the subduction of the Pacific plate under the North America plate while in the south, it is the subduction of the Philippines plate under the Eurasian plate that causes earthquakes. Japan has experienced several major and great earthquakes. Probably the most damaging earthquake in history was the 1923 Kanto earthquake of 8.3 magnitude claiming 142,800 human lives. This earthquake also caused a tsunami with a height of about 10 m. However, the damage and loss of lives was more because of shaking and the fire that broke after the earthquake. In the northern region, large earthquakes have occurred in 1611, 1896 and 1933 and every one of them produced devastating tsunamis on the Sanriku coast of Pacific NE Japan. The M 7.6 subduction earthquake of 1896 created high tsunami of 38 m and caused a reported death toll of 27,000. The M 8.6 earthquake of March 2, 1933, though not a subduction zone earthquake, produced 29 m high tsunami waves on the Sanriku coast and inundated 10 km inland along the coastal plains and claimed more than 3000 human lives.

H. K. Gupta and V. K. Gahalaut, *Three Great Tsunamis: Lisbon (1755), Sumatra–Andaman (2004) and Japan (2011)*, SpringerBriefs in Earth Sciences,
DOI: 10.1007/978-94-007-6576-4_4,

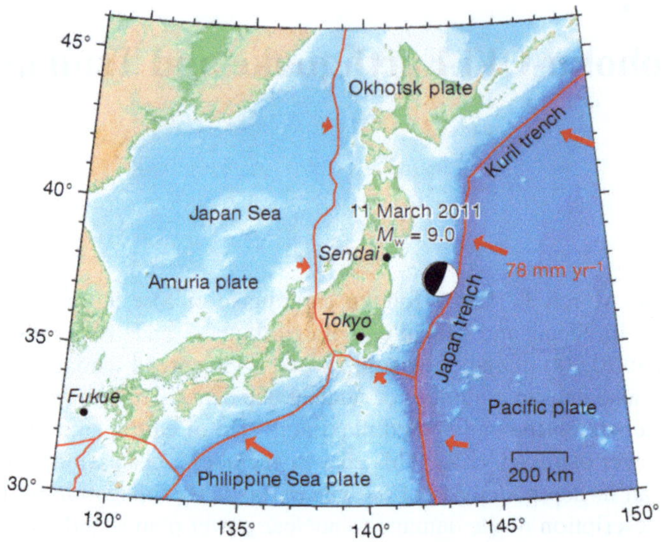

Fig. 4.1 General tectonic set-up of Japan. Location of the March 11, 2011 Tohoku earthquake is shown by the beach ball, depicting the focal mechanism of the earthquake (after Ozawa et al. 2011)

4.2 2011 Tohoku-Oki Earthquake and Tsunami

The 11 March 2011 Tohoku-Oki earthquake, which occurred at 14:46 local time (05:46 UTC), was the largest earthquake (Mw 9) in the known history of Japan (Figs. 4.2 and 4.3). The rupture of the earthquake, as estimated from the distribution of aftershocks and derived from models based on GPS, seismic waveform and tide gauge data, stretches about 300–400 km in length and 200 km in width. It generated a huge tsunami and caused 15,073 fatalities and 8,657 missing in the Tohoku and Kanto regions (Fig. 4.4). A large foreshock of 7.3 magnitude of this earthquake took place at 11:45 local time on 9 March 2011. Following the Tohoku-Oki mainshock, many aftershocks, including three with $M \geq 7.4$, occurred on the same day. Detailed analysis of the seismic waves suggests that the rupture started near the down-dip edge of the main thrust zone and propagated up-dip in both north and south directions. GPS measurements show that the coastal parts of northeastern Honshu moved up to 4 m westwards and sank by almost 1 m. Just above the earthquake hypocenter, the surface coseismic displacement was 24 m predominantly towards the east and 3 m in the up direction (Fig. 4.5). This was calculated by sea-floor geodetic observations using GPS and acoustic measurements. This was the first time that such a large coseismic displacements were measured using GPS. Such large displacements imply that the slip on the subsurface rupture must have exceeded

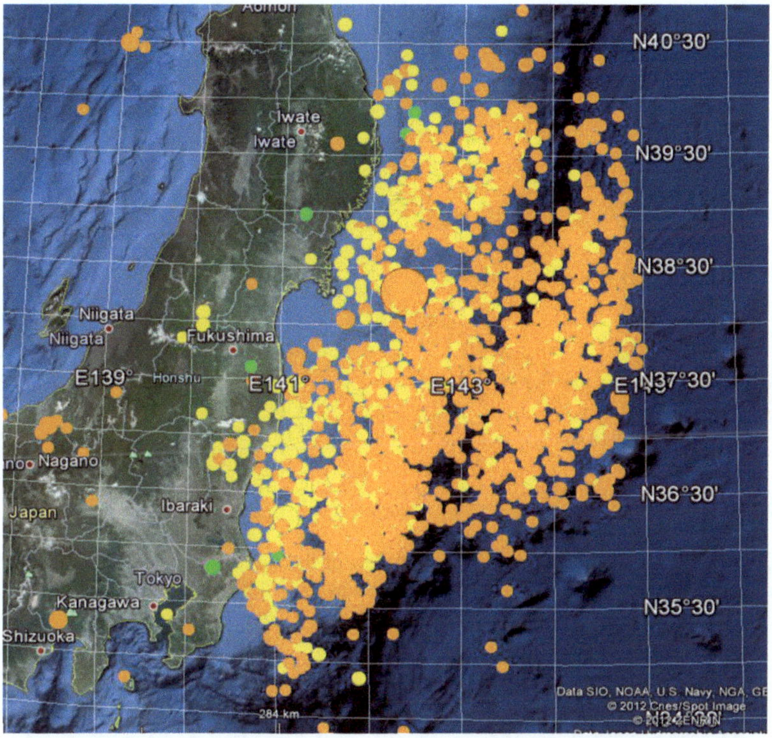

Fig. 4.2 March 11, 2011 Tohoku-Oki mainshock (the largest circle) and its aftershocks (M > 4) from USGS plotted on Google Earth. The *dark* and *light yellow* and *green colors* denote the focal depths of the aftershocks as 0–35, 35–70 and 70–150 km, respectively. The *dark blue/black color* marks the trench location

the surface displacement and when all the data are collated, the maximum slip on the rupture appears to be more than 50 m (Sato et al. 2011). This is almost double of that during the 2004 Sumatra –Andaman earthquake (Mw 9.1) and is the largest ever measured for any earthquake globally. Combined with extensive recordings from global seismic networks, the data from these seismic stations, tide gauges and GPS make the 2011 Tohoku-Oki event the best-recorded earthquake and tsunami in history. Seismic waves shook the ground in Japan with a high frequency of about 10 Hz. Ground accelerations as large as almost three times of acceleration due to gravity and peak ground velocities of 80 cm/s across Honshu were recorded. The tsunami caused by this earthquake was enormous and severely devastated the coastal regions of eastern Japan. Japan's Prime Minister Naoto Kan told reporters at a televised news conference on March 13, 2011 "*In the 65 years after the end of World War II, this is the toughest and the most difficult crisis for Japan*".

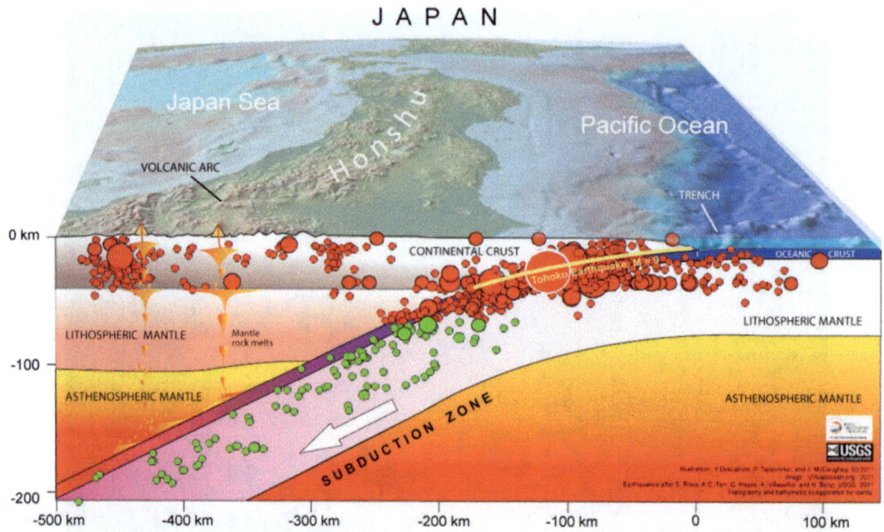

Fig. 4.3 An east–west vertical cross-sectional view of the subduction zone (USGS). *The yellow line* through the *largest circle*, showing the 2011 mainshock, depicts the earthquake rupture

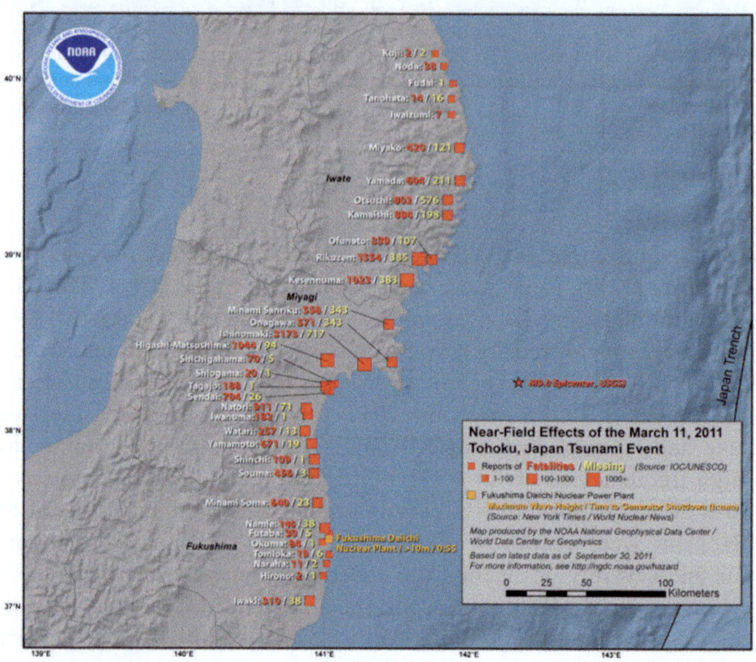

Fig. 4.4 Fatalities and missing people along the eastern coast of Japan (NOAA)

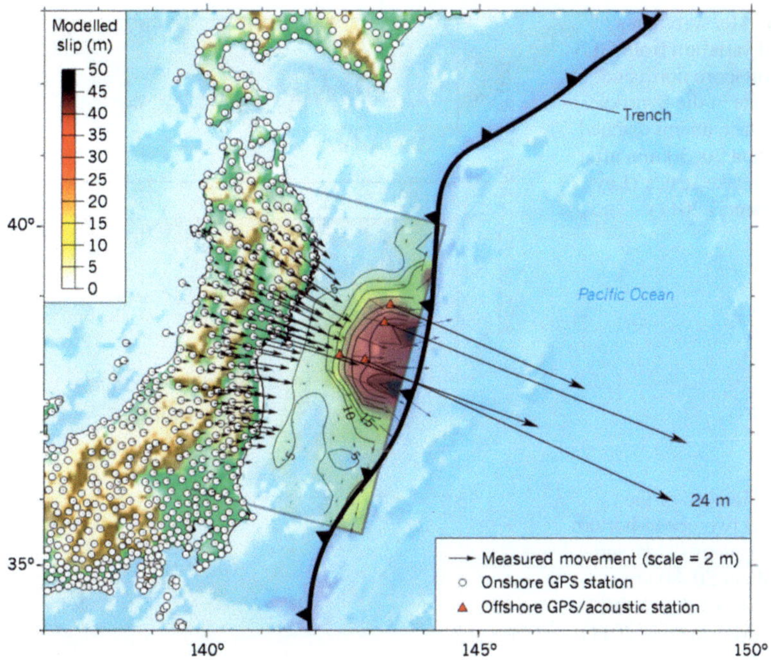

Fig. 4.5 Horizontal displacement caused by the 2011 Tohoku-Oki earthquake. The GPS sites on land are marked by *white circles* whereas the *red triangles* mark the geodetic measurements in the offshore region. The rupture model derived from these observations shows a high slip reaching 50 m (Sato et al. 2011; Newman 2011)

4.3 An Unexpected Event

The occurrence of such a large magnitude earthquake was a surprise. The observed magnitude is significantly larger than any of the earthquakes that occurred along this part of the subduction system in the past few hundred years. It appears that no earthquake larger than 8.3 had occurred earlier in this region. It had been thought that subduction of a relatively old, less buoyant oceanic lithosphere would cause an earthquake with a maximum magnitude of 8. It is now suggested that the complex plate geometry in the region had resulted in contortion of the subducting slab that actually increased plate coupling and stress build-up before the earthquake. Moreover, the shallow part of the plate interface, which was considered to be slipping aseismically, also contributed and released strain, making the width of the fault wider. This led to the extremely violent and powerful Tohoku-Oki earthquake.

Another surprise of this earthquake was the enormous tsunami it generated that swept along 70 km of the coastal plains. Several tide gauges recorded wave heights of over 4 m (Fig. 4.6) with that at Soma recording at least 7.3 m. In fact,

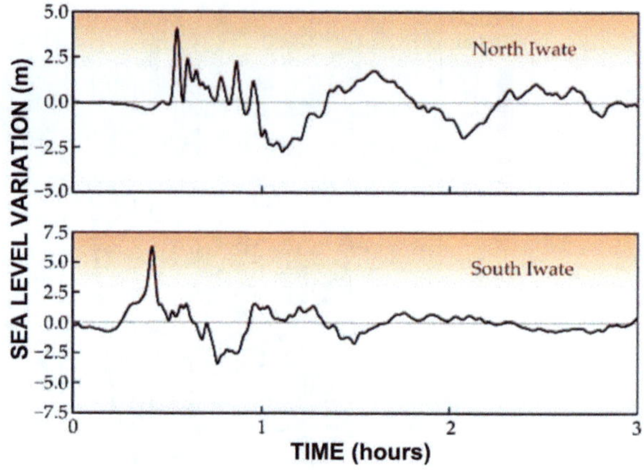

Fig. 4.6 Measurements of sea-level variation from GPS gauges offshore north and south Iwate in the Sanriku region. The tsunami reached each station 20–30 min after the earthquake struck (Lay and Kanamori 2011)

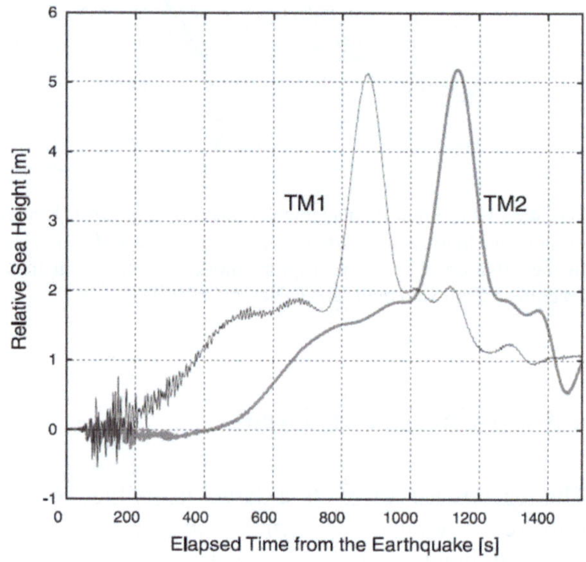

Fig. 4.7 Record of the tsunami at two ocean-bottom pressure gauges (TM1 and TM2), about 50–80 km off the coast of Kamaishi, Sanriku (Maeda et al. 2011)

many tide gauges became saturated and the amplitude got clipped. Two ocean-bottom pressure gauges, which are located about 50–80 km off the coast of Kamaishi, Sanriku, recorded a 5 m high tsunami (Maeda et al. 2011) (Fig. 4.7). Despite much of the coast being protected by tsunami walls, these were not designed to stop 10–15 m high waves that inundated the coast. In Sendai city, the area inundated by the tsunami was almost 5 km inland, whereas official maps indicated only about 1 km of tsunami evacuation area from the coast (Fig. 4.8). Several towns, with houses built using timber frames designed to be flexible to withstand earthquake shaking, were simply swept away. Even the tree line that was planted on the coast could not be of any use in arresting the fury of the tsunami. In fact the uprooted trees, which were swept along with the tsunami waves, added to the force of these waves and caused more damage than providing any protection.

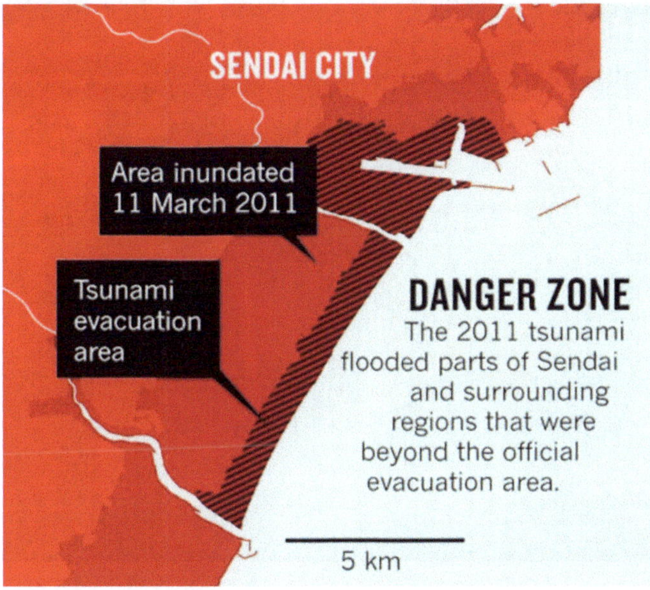

Fig. 4.8 Part of the map of Sendai city showing the official tsunami evacuation area and the actual area that was inundated during the 2011 Tohoku-Oki earthquake (Cyranoski 2012)

The tsunami waves inundated more than 500 km^2 of land across six prefectures, destroying nearly 130,000 buildings and damaging 245,000 others (Figs. 4.9 and 4.10). About 240,000 cars were washed away and destroyed. About fifteen thousand people are known to have been killed and about three thousand people are still missing. Estimates of the damage in this region range from US$14.5 to 34.6 billion. The World Bank estimated the total economic cost as US$235 billion, thus becoming the world's most expensive natural disaster.

In Japan, earthquake monitoring systems have been deployed quite extensively. Building codes have been strictly implemented. These two factors undoubtedly saved lives. Moreover, as tsunami waves travel slowly in shallow coastal water, many people were able to reach high ground in time to escape the flooding. However, some reports indicate that many people did not take immediate action out of a belief that the extensive network of tsunami walls would protect them. Given a more accurate early estimate of the true enormity of the event, $Mw = 9$, the JMA would have been able to issue a warning that might have prompted more extensive evacuations (Lay and Kanamori 2011). In Tokyo, the early warning system issued an alert one minute before the earthquake waves arrived as data from seismometers close to the epicenter could be transmitted and processed faster than the seismic waves could travel. But the initial 5–10 s of shaking produced by the Tohoku-Oki earthquake was weak, comparable to a magnitude 4.9 event and the JMA's earthquake early warning system underestimated the expected overall intensity (Lay and Kanamori 2011). Nevertheless, due to the availability of other early warning systems, more than 20 high-speed bullet trains could be stopped in the Tohoku district when the earthquake struck.

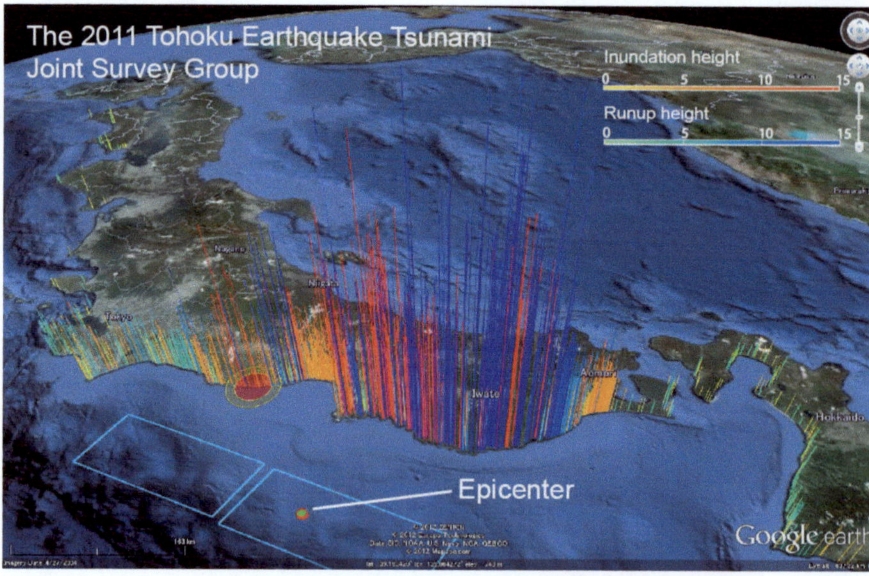

Fig. 4.9 Inundation and run-up heights along the eastern coast of Japan caused by the tsunami (Tsunami Joint Survey group http://www.coastal.jp/tsunami2011/). The source zone of the earthquake is shown by the two rectangles

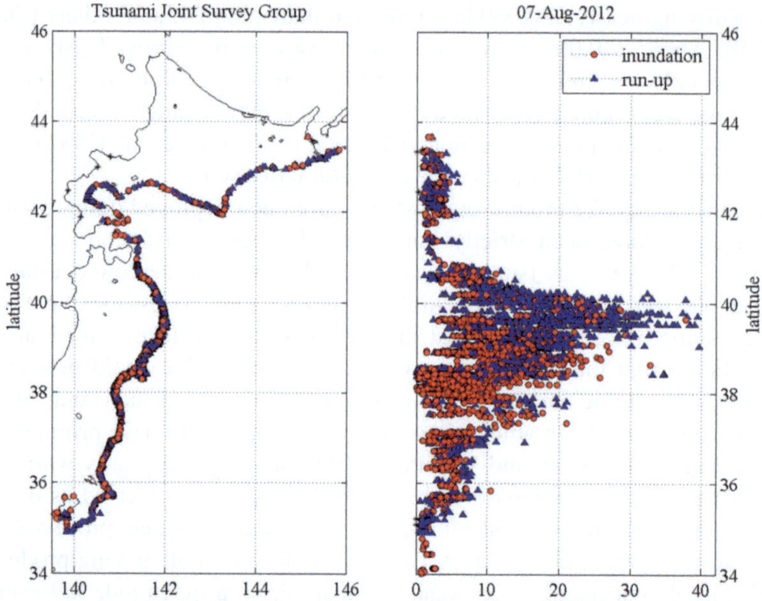

Fig. 4.10 Inundation and run-up heights along the eastern coast of Japan caused by the tsunami (Tsunami Joint Survey group http://www.coastal.jp/tsunami2011/)

4.4 Effect of the 2011 Tsunami in Japan

The USGS initially estimated the size of the earthquake as 7.9 and then upgraded it to 8.8 and then quickly to 8.9 and finally to 9.0. The Pacific Tsunami Warning Centre issued its first preliminary regional tsunami warning ten minutes after the earthquake struck, upgrading it to a widespread tsunami warning after another hour and a half. The tsunami warning issued by the Japan Meteorological Agency, after only 2 min and 40 s, was the most serious with its warning scale and rated it as a "major tsunami" and advised that 6 m, 3 m and 3 m tsunamis could be expected along the coast of the Miyagi, Iwate and Fukushima prefectures, respectively. Initial estimates indicated that the tsunami would reach in 10–30 min the areas first affected and then the areas farther north and south, based on the geography of the coastline, would be affected. However, the actual tsunami caused by the earthquake was much more severe and an underestimated forecast led to slow evacuation. Also, many people caught in the tsunami thought that they were located on high enough ground to be safe. The tsunami seawalls at several of the affected cities were based on much smaller estimated tsunami heights. Thus, among several factors causing the high death toll from the tsunami, one was the unexpectedly large size of the water surge. Just over an hour after the earthquake at 15:55 JST, a tsunami was observed flooding Sendai Airport, which is located near the coast of Miyagi Prefecture, with waves sweeping away cars and planes and flooding various buildings as they traveled inland. A 4 m high tsunami hit Iwate Prefecture. Wakabayashi Ward in Sendai was also particularly hard hit (Figs. 4.11, 4.12, 4.13, 4.14, 4.15, 4.16, 4.17, 4.18, 4.19, 4.20, 4.21).

The Japan Meteorological Agency (JMA) published details of tsunami observations recorded around the coastline of Japan following the earthquake. The timing of the earliest recorded tsunami maximum readings ranged from 15 h 12 min to 15 h 21 min, that is 26 and 35 min after the earthquake had struck. The bulletin also included initial tsunami observation details, as well as more detailed maps for the coastlines affected by the tsunami waves. These observations included maximum tsunami readings of over 3 m at the following locations:

Arrival time	Place	Height of the tsunami (m)
15:12 JST	Off Kamaishi	6.8
15:15 JST	Ōfunato	>3.2
15:20 JST	Ishinomaki-shi Ayukawa	>3.3
15:21 JST	Miyako	>4.0
15:21 JST	Kamaishi	>4.1
15:44 JST	Erimo-cho Shoya	3.5
15:50 JST	Sōma	7.3
16:52 JST	Ōarai	4.2

At a few places, the tsunami height was inferred to be more than 30 m. A joint research team from Yokohama National University and the University of Tokyo reported that the tsunami at Ryōri Bay, Ōfunato was about 30 m high. At Tarō, Iwate, a University of Tokyo researcher reported an estimated tsunami height of 37.9 m, which reached the slope of a mountain some 200 m away from the coastline.

Fig. 4.11 People watching the tsunami at Sendai (H Kawahara/AFP/Getty)

Fig. 4.12 Sendai airport on March 16, 2011 flooded with water inundation due to the tsunami

4.5 Tsunami Across the Pacific

Immediately after the warning by the Pacific Tsunami Warning Center (PTWC) in Hawaii, evacuation around the Pacific Ocean started where tsunami waves reached more than 6 h after the earthquake (Fig. 4.22). Russia evacuated 11,000 residents from coastal areas of the Kuril Islands. The United States West Coast and Alaska Tsunami Warning Center issued a tsunami warning for the coastal areas in most

Fig. 4.13 Smoke rises from a burning factory in Sendai, March 12. (Kyodo/Reuters)

Fig. 4.14 Tsunami debris scattered over a devastated area of Sendai, March 14, 2011. (STR/AFP/Getty Images)

of California, all of Oregon and the western part of Alaska and a tsunami advisory covering the Pacific coastlines of most of Alaska and all of Washington and British Columbia, Canada. In California and Oregon, up to 2.4 m high tsunamis hit some areas, damaging docks and harbors and causing over US$10 million in damage. A tsunami of up to 1 m hit Vancouver Island in Canada prompting some evacuations and a ban of boats in the water surrounding the island for 12 h. In

Fig. 4.15 Vehicles pass through the ruins of the leveled city of Minamisanriku, northeastern Japan, Tuesday March 15, 2011. (AP Photo/David Guttenfelder)

Fig. 4.16 A truck dangles from a collapsed bridge in Ishinomaki, northern Japan, four days after the earthquake (AP Photo/The Yomiuri Shimbun, Hiroshi Adachi)

Fig. 4.17 People walk a road between the rubble of destroyed buildings in Minamisanriku town, Miyagi Prefecture, northern Japan, three days after the earthquake (AP Photo/The Yomiuri Shimbun, Tsuyoshi Matsumoto)

Fig. 4.18 A fishing boat rests surrounded by debris in the city of Kamaishi, Iwate Prefecture on March 12. (Yomiuri Shimbun/AFP/Getty Images)

Fig. 4.19 Buildings are covered with mud in Minamisanriku, Miyagi Prefecture, March 12. (Naoki Ueda/The Yomiuri Shimbum/Associated Press)

the Philippines, waves of up to 0.5 m high hit the eastern seaboard of the country. Authorities in Wewak, East Sepik, Papua New Guinea evacuated 100 patients from the city's Boram Hospital before it was hit by waves, causing an estimated US$4 million of damage. In Hawaii, the estimated damage to public infrastructure was US$3 million. It was reported that a 1.5 m high wave completely submerged Midway Atoll's reef inlets and Spit Island, killing more than 110,000 nesting seabirds at the Midway Atoll National Wildlife Refuge. Some other South Pacific countries, including Tonga and New Zealand and U.S. territories including American Samoa and Guam, experienced larger-than-normal waves but did not report any major damage. However, in Guam, some roads were closed and people were evacuated from low-lying areas. Along the Pacific Coast of Mexico and South America, tsunami surges were reported but in most places caused little or no damage. Peru reported a wave of 1.5 m and more than 300 homes damaged. The tsunami in Chile was large enough to damage more than 200 houses, with waves up to 3 m in height.

The wave height was generally a few tens of cm as the tsunami crossed the Pacific but increased as it reached shallow coastal waters, with waves up to 3 m arriving on the coast of Chile about 20–21 h after the earthquake. One man was killed in Indonesia and another died in California attempting to photograph the waves. The impact on the coast of the United States was lessened as the arrival of the tsunami waves largely coincided with low tide but tens of millions of dollars' worth of damage was inflicted to ports and harbors. Damage to houses was also reported from Peru, Chile and Indonesia.

Fig. 4.20 People walk on debris scattered across the town of Minamisanriku, Miyagi Prefecture on March 12. (Yomiuri Shimbun/AFP/Getty Images)

4.6 Fukushima Meltdown

The unprecedented height of the tsunami waves not only damaged property and killed several thousand people, it also caused another scare, a nuclear accident. This accident is the second biggest after the Chernobyl nuclear disaster of 1986 but more complex as all eleven reactors were involved. Japan produces about 1000 TWh of electricity and about one fourth of this is met by 53 nuclear reactors. Japan had 282 GW of total installed electricity generating capacity in 2010. However, after the damage by the 2011 earthquake, the capacity was estimated to be around 243 GW in mid-2011.

Fig. 4.21 A home drifts in the Pacific Ocean on March 13, 2011 (REUTERS/U.S. Navy/Mass Communication Specialist 3rd Class Dylan McCord)

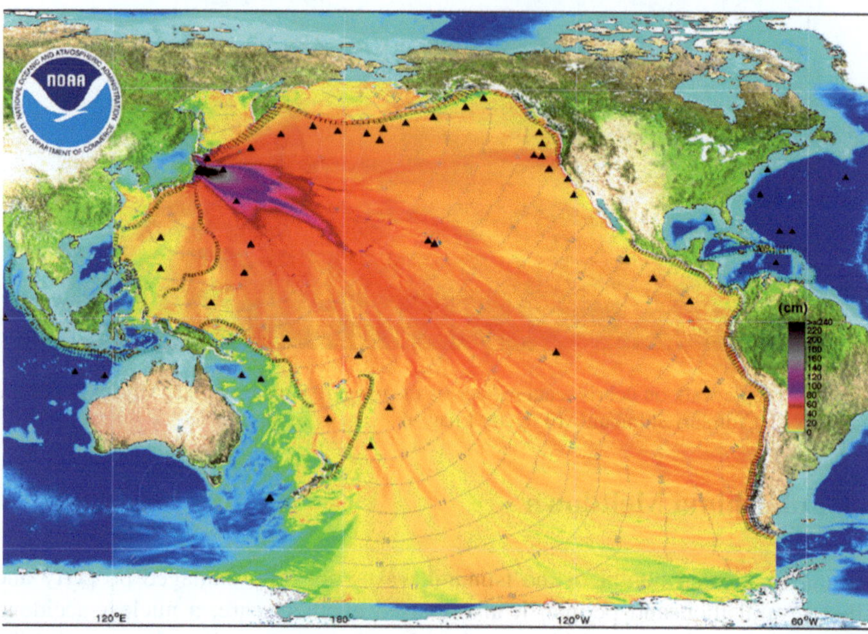

Fig. 4.22 Simulated tsunami amplitude in the open ocean for the 2011 Tohoku-Oki earthquake (National Oceanic and Atmospheric Administration, NOAA, USA). The *grey color* contours show the traveltimes of the tsunami waves

The Fukushima Daiichi, Fukushima Daini, Onagawa Nuclear Power Plant and Tōkai nuclear power stations, consisting of eleven reactors, were automatically shut down following the earthquake. Cooling is needed to remove decay heat after a reactor has been shut down and to maintain spent fuel pools. The back-up cooling process is powered by emergency diesel generators at the plants and at the Rokkasho nuclear reprocessing plant. At Fukushima Daiichi and Daini, tsunami waves overtopped seawalls and destroyed diesel back-up power systems. Three large explosions and radioactive leakage occurred (Fig. 4.23). Japan declared a state of emergency, following the failure of the cooling system at the Fukushima Daiichi nuclear power plant, resulting in the evacuation of nearby residents. Over 200,000 people were evacuated. Officials from the Japanese Nuclear and Industrial Safety Agency reported that radiation levels inside the plant were up to 1,000 times more than normal levels and that radiation levels outside the plant were up to 8 times more than normal levels. Later, a state of emergency was also declared at the Fukushima Daini nuclear power plant located about 11 km south. Radioactive iodine was detected in the tap water in Fukushima, Tochigi, Gunma, Tokyo, Chiba, Saitama and Niigata and radioactive cesium in the tap water in Fukushima, Tochigi and Gunma. Radioactive cesium, iodine and strontium were also detected in the soil in some places in Fukushima. Food products were also found contaminated by radioactive matter in several places in Japan. On 5 April 2011, the government of Ibaraki Prefecture banned the fishing of *sand lance* (or sand eels, a variety of fish) after discovering that this species was contaminated by radioactive cesium above legal limits. Only after a few years will we know the actual damage done by the radiation to living beings and the ecosystem.

4.7 Lessons Learnt from the 2011 Earthquake

The amount of data generated by this earthquake is unprecedented. These data will be used to understand the occurrence of giant earthquakes, their potential of generating tsunamis and causing damage. Insights gained from the Tohoku earthquake are helpful to scientists to re-evaluate the seismic hazard. This will contribute to improved scenario building, code development and public warnings about tsunami threats. It is expected that these studies will help in mitigating the hazards due to future large earthquakes. One thing that seismologists have now learnt is the possibility of occurrence of giant earthquakes along subduction zones. Japanese scientists had not estimated that an earthquake of such a large magnitude could occur in that area. The tsunami seawalls in the area were built for a tsunami resulting from a magnitude 8.0 earthquake and not a 9.0 magnitude earthquake. Thus, even though the Japanese had planned and were well-prepared for an M 8 earthquake, which might have a recurrence interval of 200–300 years, they were not prepared for a giant earthquake with a recurrence interval of, say 1000 years. Another issue worth mentioning here is that Japan was focusing more on the Tokai and Nankai regions (south of Tokyo) where they expected large earthquakes to occur. Thus the

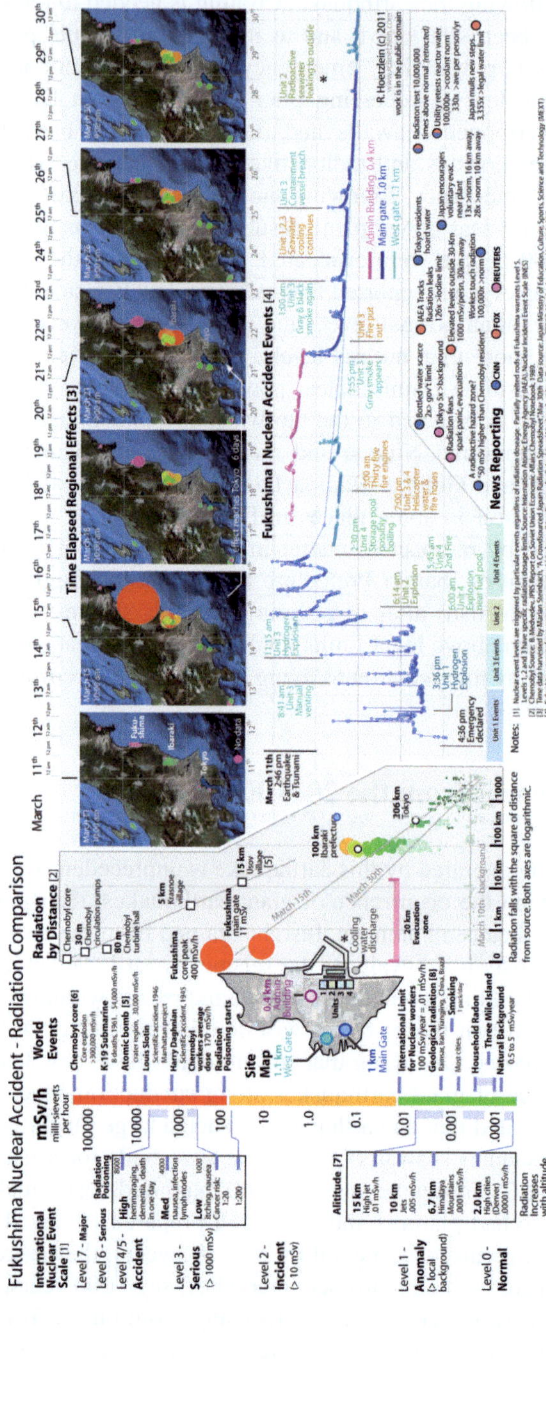

Fig. 4.23 Severity of the Fukushima nuclear accident (*source* Wikipedia)

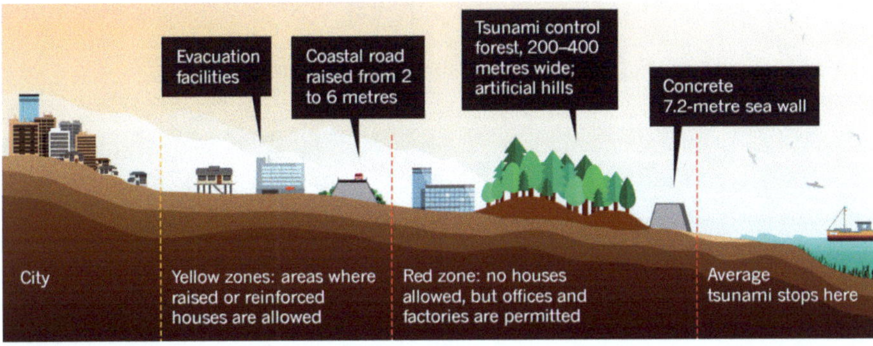

Fig. 4.24 A new proposed plan for Sendai city (Cyranoski 2012)

projected seismic hazard in those regions was higher than in the region hit by the 2011 earthquake (Geller 2011). Consequently, Japan is currently updating its tsunami disaster plans for all of its coastal areas (Fig. 4.24). It has been suggested that all plans take evidence from paleo-tsunami deposits into consideration. Even relatively long seismological records are too limited to adequately assess the hazard from infrequent but devastating events. From a recent re-evaluation of palaeo-tsunami deposits, more than three kilometers inland of the Sendai plain, it is now inferred that an earthquake of magnitude ~9 did occur in the same region of Japan in 869 (Minoura et al. 2001).

References

Cyranoski D (2012) After the deluge. Nature 483:141–143

Geller R (2011) Shake-up for Japanese seismology. Nature 472:407–409

Lay T, Kanamori H (2011) Insights from the 2011 Japan earthquake. Phys Today, pp 33–39

Maeda T, Takashi F, Shinichi S, Masanao S (2011) Significant tsunami observed at ocean-bottom pressure gauges during the 2011 off the Pacific coast of Tohoku earthquake. Earth Planets Space 63:803–808

Minoura K, Imamura F, Sugawara D, Kono Y, Iwashita T (2001) The 869 Jogan tsunami deposit and recurrence of large-scale tsunami on the coast of Northeast Japan. J Nat Disaster Sci 23:83–88

Newman AV (2011) Hidden depths. Nature 474:441–443

Ozawa S, Takuya N, Hisashi S, Tomokazu K, Mikio T, Tetsuro I (2011) Coseismic and postseismic slip of the 2011 magnitude-9 Tohoku-Oki earthquake. Nature 475:373–376

Sato M, Saito H, Ishikawa T, Matsumoto Y, Fujita M, Mochizuki M, Asada A (2011) Restoration of interplate locking after the 2005 Off-Miyagi prefecture earthquake, detected by GPS/acoustic seafloor geodetic observation. Geophys Res Lett 38: (L01312) doi:10.1029/2010GL045689

Chapter 5
Global Effort to Forecast and Mitigate a Tsunami Hazard

Abstract This chapter gives a description of the key components of tsunami warning centers, modeling scenario data bases, inundation modeling, types of tsunami advisories and global tsunami warning centers.

5.1 Introduction

Tsunamis are primarily caused by earthquakes. As of now, it is not possible to forecast earthquakes. However, soon after an earthquake occurs, its parameters (origin time, location and magnitude) are estimated. It is then possible to estimate whether a particular earthquake would generate a tsunami, how strong this tsunami would be and what its size and arrival time would be at various locations locally and globally. The first scientific effort in this direction dates back to the 1920s when an attempt to warn about tsunamis was made in Hawaii. Later, responding to the 1st April, 1946 Aleutian Island earthquake and the May 23, 1960 Valdivia earthquake and the resultant tsunamis that caused massive destruction in Hilo, Hawaii, more advanced tsunami warning systems were developed. The basics of developing tsunami warning systems is the fact that in the open ocean tsunamis travel with a speed of 500–1000 km/h, whereas the longitudinal seismic waves travel much faster at a speed of 5–7 km/s (18,000–25200 km/hr). So an earthquake could be located within a few minutes of its occurrence, while the tsunami waves will take several minutes to several hours to reach a vulnerable location, depending upon its distance from the earthquake source.

From the point of view of early warning of tsunamis, they are classified into the following three categories by the Inter-Governmental Commission on Oceanography (UNESCO-IOC 2006; Nayak and Kumar 2011):

Local Tsunami: These are caused by a source located close to the coast (within 100 km, or less than 1 h of the tsunami traveltime to the coast) and are the most

H. K. Gupta and V. K. Gahalaut, *Three Great Tsunamis: Lisbon (1755), Sumatra–Andaman (2004) and Japan (2011)*, SpringerBriefs in Earth Sciences,
DOI: 10.1007/978-94-007-6576-4_5, © Springer Science+Business Media Dordrecht 2013

destructive. There is very little time to react, evacuate and implement other defensive measures.

Regional Tsunami: These are capable of creating destruction in a particular geographic region within a distance of 1000 km from the source and the tsunami traveltime could be 1–3 h.

Distant Tsunami: A distant tsunami is caused by a major earthquake in a subduction zone and is capable of causing destruction in the Pacific Rim countries located around the ocean basin to distances of more than 1000 km and a tsunami traveltime of more than 3 h.

5.2 Components of Tsunami Warning Centers

5.2.1 Seismic Network

A tsunami warning center should be able to locate an earthquake within a few minutes of its occurrence. This requires the availability of earthquake data from seismic stations distributed globally. If the tsunami warning center is catering to the needs of a specific region, availability of a dense regional network of seismic stations is very desirable. For accurate estimate of the moment-magnitude of an earthquake, it is necessary to have seismic stations with broadband, low noise and high-dynamic range digital seismic data acquisition capabilities. Timely estimation of earthquake parameters, say within 5 min is desirable. The density of seismic stations should be such that there are several (8–10) seismic stations within 900 km of the earthquake source.

5.2.2 Sea-Level Network

To determine whether an earthquake has generated a tsunami, it is necessary to monitor changes in the water level as close to the source as possible. For this purpose tide gauges, ocean- bottom pressure recorders and coastal ocean dynamics application radars are used. These devices provide accurate sea-level data in real or near real time to determine whether an earthquake has generated a tsunami and if so, how big it will be when it reaches the coasts of the likely affected countries. There are a number of global networks that provide sea-level information in real time and they are coordinated by the Inter-Governmental Commission of Oceanography (IOC) under the UNESCO umbrella. The Global Sea Level Observing System (GLOSS), operated under the guidance of the Joint Technical Commission for Oceanography and Marine Meteorology (JCOMM) of the World Meteorological Organization (WMO) and IOC, is the most prominent among these networks.

5.2.3 Modeling

Numerical modeling is used to estimate the mechanism of tsunami generation from earthquake data in addition to forecasting the traveltime of tsunami waves to different locations, identifying the inundation areas and extent of inundation. These are all important components of early tsunami warning systems. A number of tsunami models are available with relative advantages and disadvantages. The most commonly used models are the TUNAMI (Tohoku University Numerical Analysis Model) developed by Imamura et al. (1995) and the MOST (Method of Splitting Tsunami) pioneered by Titov and Gonzalez (1997). Using the data of global seismic stations, these models make use of earthquake focal parameters like location and depth, strike, dip and rake as well as the length and width of the rupture plane and slip to compute the generation and propagation of tsunamis locally and of the trans-ocean basin as well as inundation.

5.2.4 Scenarios

Running of the tsunami models after the occurrence of an earthquake takes a lot of time, which can be avoided by preparing a pre-run data base. It is very helpful to have a look-up table (LUT) available for the possible scenarios. This has been achieved by using unit source function methodology, where each unit is equivalent to a tsunami generated by an Mw 7.5 earthquake having a rectangular source of 100 km by 50 km and a 1 m slip. These units are suitably combined to produce the tsunami scenario for a relevant earthquake (Gica et al. 2008; Greenslade and Titov 2008; Nayak and Kumar 2008a, b, c). The entire circum-Pacific seismic belt, the Caribbean (for the Atlantic region) and the Indian Ocean have been modeled with such an approach. To match an earthquake after its occurrence, the basic unit source scenarios are selected, merged and scaled up/down based on scaling relations to generate a scenario corresponding to the earthquake under consideration (Nayak and Kumar 2008c).

5.2.5 Inundation

An important issue in mitigating a tsunami hazard is the knowledge of the possible inundation and therefore the extent of the human population likely to be affected as well as the major civil structures and the facilities that are likely to be damaged by the tsunami. For developing tsunami run-up and inundation scenarios, it is necessary to have close grid information of shallow bathymetry and near-shore topography. The height of the tsunami wave, as it approaches the coast, is governed by the near-coast bathymetry. Similarly, how a tsunami wave would inundate a coastal region is governed by the near-coast topography. In the tsunami prone coastal areas, detailed near-shore bathymetric and topographic surveys are carried out and estimates are made for possible inundation for a given height of a tsunami.

5.2.6 Communication

Having found that an earthquake has occurred and it is likely to generate a tsunami that would affect a given coastal region, it is extremely important to communicate an estimate of the severity and the time of arrival to the likely effected population in the minimum possible time. This requires a reliable communication infrastructure. The tsunami warning centers use multiple lines of communication such as radios, SMS, e-mail, television, texting and telex etc. Particularly, the emergency service operators need to be informed of the pending tsunami. As mentioned earlier, geographically there are several inhabited coastal areas (several locations in Japan, Indonesia and elsewhere) where within a few minutes of the occurrence of an earthquake, the tsunami would inundate the coast. For such areas, a tsunami advisory is issued as soon as an earthquake of M 6 or more occurs in the seismic zone prone to generating tsunamis.

For issuing a tsunami warning, decision support systems have been developed. These are basically coded standard operation procedures that compile earthquake data in real time and combine them with tsunami scenario models to accurately generate tsunami warning advisories that need to be sent to areas of concern as early as possible.

5.3 Tsunami Warning Centers

From the early installation of the rudimentary tsunami alert system of the 1920s, today, there are several tsunami warning facilities deployed all over the world. Seven of these have independent services. These are the Pacific Tsunami Warning Center (PTWC), Japan Meteorological Agency's (JMA) center, Sakhalin Tsunami Warning Center (STWC), German Indonesian Tsunami Early Warning Center (GITEWS), Joint Australian Tsunami Warning Center (JATWC) and the Indian Tsunami Early Warning Center (ITEWS). These centers along with the areas covered by them are shown in Fig. 5.1 (after Nayak and Kumar 2011).

5.4 Tsunami Advisories

Over the years, the science of tsunami warning has significantly advanced. It has become very important that appropriate information is provided to the officials involved with handling disaster related issues as well as to the general public. The advisories have to be in simple language and should include the necessary information, so that the appropriate steps are taken by the users. The first advisory is on the basis of the occurrence of an earthquake and is qualitative. All other advisories, following the first earthquake advisory, are quantitative. Following is the gist of the advisories as issued by the global tsunami warning centers (after Nayak and Kumar 2011):

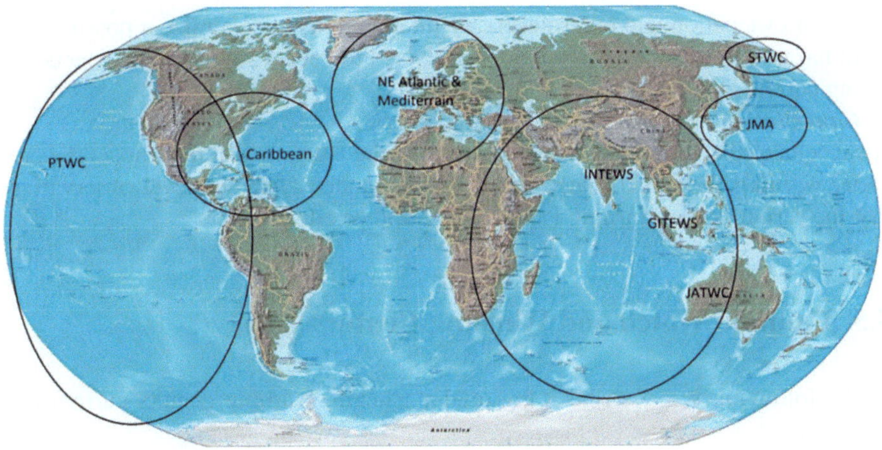

Fig. 5.1 Global tsunami warning services and the areas covered by them (after Nayak and Kumar 2011). For details see text

Earthquake information bulletin

The bulletin contains the focal parameters of the earthquake such as the origin time, latitude and longitude of the epicenter and the focal depth, magnitude and geographical location where it has occurred. This bulletin also comments on the potential of causing a tsunami depending upon the geographical location and the magnitude of the earthquake.

Warning

When the potential of a tsunami is high and a large coastal area is likely to be affected, the highest level of the advisory, the warning, is issued. The warning also alerts the public, likely to be affected by the ensuing tsunami, of coastal flooding, the anticipated height of the tsunami when it arrives as well as the expected time of its arrival at various coastal locations. With the passage of time the warnings could be updated; the geographical location to be affected can be suitably adjusted based on the information collected; downgraded or cancelled.

Watch

When immediate public evacuation is not required, a watch advisory is issued to alert the emergency management officials. The watch advisories are based on seismic observations of a large enough earthquake that has occurred in a region where a tsunami could be generated, without the confirmation of the generation of a tsunami. Officials should be ready to effect evacuation, in case the watch is upgraded to a warning. Later, the watch may be cancelled.

Alert

It is a lower category of 'Watch', issued on the basis of an earthquake occurrence, with a qualitative assessment of a possible tsunami. With the passage of time, with more information gathered, the 'Alert' could be upgraded to a 'Watch' or 'Warning' or cancelled. Local disaster management authorities are expected to be ready for evacuation and other necessary interventions to minimize a possible tsunami hazard. Under alert, the public are advised to keep away from the beaches as strong currents are expected.

Cancellation

A cancellation is issued after ascertaining that the tsunami will not impact the area under warning. It is also issued when it is estimated from sea-level observations that the earlier issued warning is no longer valid or effective.

5.5 Tsunami Watch and Warning Centers

In the following we describe a few major tsunami warning centers currently in operation.

5.5.1 The Pacific Tsunami Warning Center

This is the oldest tsunami warning center set up in 1949 in response to the 1946 Aleutian Island earthquake and the resultant tsunami that claimed 165 human lives in Alaska and Hawaii, besides colossal damage to property. Located at Ewa Beach, Hawaii, it was originally known as the Honolulu Observatory. The 1960 Chilean earthquake caused an ocean-wide tsunami across the entire Pacific Ocean basin. Consequently, the scope of the Honolulu Observatory was enhanced to provide tsunami information for the entire Pacific Rim countries and it was re-named the Pacific Tsunami Warning Center (PTWC). It provides warning and advisories to participating members and nations in the Pacific Ocean area. It also acts as the local warning center for the state of Hawaii. Following the devastating Sumatra earthquake and the deadliest tsunami so far in 2004, the scope of PTWC was further expanded to include the Indian Ocean and the Caribbean Sea countries.

5.5.2 The West Coast and Alaska Tsunami Warning Center

There was a massive Mw 9.4 earthquake on 27th March, 1964 in Alaska that underlined the necessity to create facilities to provide timely and effective earthquake and tsunami warnings to the coastal areas of Alaska. This resulted in establishing the Palmer Observatory in 1967 at Palmer city in Alaska and it was known as the Alaska Regional Tsunami Warning System (ARTWS). It served all the coastal regions of USA and Canada. In 1973 the ownership of ARTWS changed and it was renamed as the Alaska Tsunami Warning Center (ATWC). In 1996, its responsibility was expanded to include the coastal areas of California, Oregon, Washington, British Colombia and Alaska and the present name West Coast/Alaska Tsunami Warning Center (WC/ATWC) was adopted.

5.5.3 The Japan Meteorological Agency

Japan is one of the most earthquake and tsunami prone countries in the world. The March 11, 2011 earthquake of magnitude Mw 9.0 and the resultant tsunami caused wide spread damage. The nuclear power plants in Fukushima suffered considerable damage. About 20,000 human lives were lost. The meteorological services in Japan were initiated in 1875. Currently, known as the Japan Meteorological Agency (JMA), under the Ministry of Land, Infrastructure, Transport and Tourism (MLIT), it is responsible for prevention and mitigation of natural disasters. In addition to serving the Japanese Islands, it also provides advisories on a global scale. After the occurrence of an earthquake, JMA estimates the possibility of a tsunami being generated on the basis of seismic data. In case a tsunami is expected in the coastal regions of Japan, JMA issues tsunami warning/advisories within 2–3 min of the occurrence of the earthquake. For far-off tsunamis, JMA co-operates with PTWC in issuing appropriate warning/advisories globally.

5.5.4 Sakhalin Tsunami Warning Center

In response to the devastating tsunami generated by the November 4, 1952 Kurile (M 9.0) earthquake, Russia set up the Sakhalin Tsunami warning Center (STWC) in 1958. Located in Yunzo-Sakhalinsk it has two components, namely the Tsunami Warning Center (TWC) and the Seismic Station of the Russian Academy of Sciences. The primary aim of STWC is to issue tsunami warning/advisories to the Civil Defense and Emergency Regional Headquarter and to the Central Telegraph Station of Yuzano-Sakhalinsk about possible tsunamis from near and far-off earthquake sources.

5.5.5 German-Indonesian Tsunami Early Warning System

Indonesia is very vulnerable to tsunamis. The earthquakes occurring along the subduction zone in the Sunda Trench are very close to major Indonesian cities and a tsunami could reach within a short time of 10–20 min, providing very little time to affect defensive measures and cause evacuation. This limiting factor was the basis of setting up GITEWS. Supported by the Government of Germany, GITEWS was completed and handed over to Indonesia in March, 2011. It has demonstrated its effective operation since then by effectively giving tsunami advisories within 5 min of the occurrence of earthquakes. GITEWS is based in Jakarta at the Indonesian Meteorological, Climatological and Geophysical Agency. The system is classified as a modern tsunami warning system globally.

5.5.6 Joint Australian Tsunami Warning Center

Based in Melbourne and Canberra, the Joint Australian Tsunami Warning Center (JATWC) is operated by the Bureau of Meteorology and Geo-science Australia (GA). The purpose is to provide Australia an independent capability to detect, monitor, verify and provide appropriate advisories to Australian coastal regions about tsunamis.

5.5.7 Indian Tsunami Early Warning System

Soon after the 2004 tsunami, India took up the work of establishing a modern tsunami warning center and the Indian Tsunami Early Warning System (ITEWS) came into operation in August 2007. Since then it has been operating uninterrupted and is capable of giving accurate tsunami advisories within 10 min of the occurrence of an undersea earthquake of magnitude 6.5 or larger occurring anywhere in the Indian Ocean. Kumar et al. (2012a) have analyzed the performance of ITEWS over the past 5 years and found it to perform very well. Here we give some details of the components and the operation procedure adopted by this very successful facility.

Observation System of ITEWS

The system receives seismic data from about 300 seismic stations globally in real time. The focal parameters of the earthquakes are determined using autolocation software. The sea-level observations are made using ocean-bottom pressure recorders and coastal tide gauges. It must be mentioned here that in the Indian Ocean, it has been found that there are only two known sources capable of generating tsunamigenic earthquakes. The first is the seismic belt between Sumatra and Andaman Islands, a stretch of some 4000 km in length. The second source is the area off the Makaran coast in the Arabian Sea (Fig. 5.2). This is important, particularly for the coastal regions of India. It must be mentioned here that tsunamis are rather rare for the Indian coastal region. In the entire twentieth century, there were only three tsunamis to have indented the east coast of India. The 26th December 2004 tsunami caused immense loss of human lives. Many of them lost their lives because of ignorance. After the first wave hit the Tamil Nadu coast line, the sea withdrew. It was a Sunday morning and the local time was around 9 am and people were walking on the seashore. After the sea withdrew, people just walked in out of curiosity or to pick sea shells, they did not know what was in store for them. The second tsunami wave claimed lives of the morning walkers who were unknowingly trapped. It is very important to reduce the number of false tsunami warning alarms, as happened on 28th March, 2005. The Nias earthquake of Mw 8.7 occurred very close to the epicenter of the 26th December, 2004 earthquake at 19.09 pm Indian Time. A tsunami was forecasted and evacuation was affected. There was no significant tsunami. However, the tsunami warning was withdrawn early in the morning of 29th March, 2005. This caused immense inconvenience to millions of evacuees. Fortunately, other than the Andaman and Nicobar group of islands, the other coastal areas are

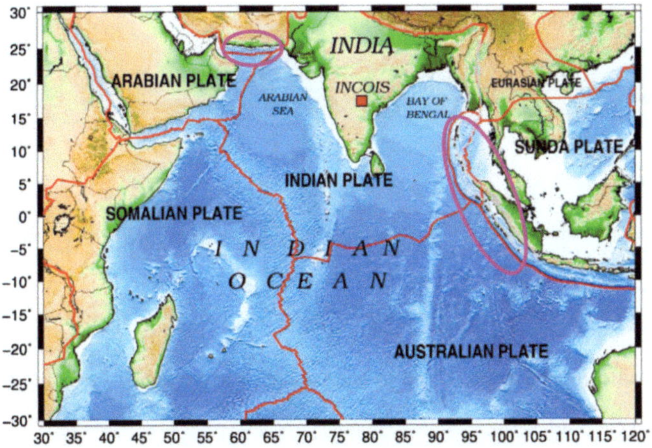

Fig. 5.2 Regional tectonic setting of the Indian Ocean. The location of ITEWS at INCOIS, Hyderabad is shown by a red square in the map of India. Two source regions capable of generating tsunamigenic earthquakes (Makaran and Sumatra–Andaman regions) are shown by pink ellipses (modified after Kumar et al. 2012a)

2 h or more traveltime away from the two tsunamigenic earthquake sources for India. To reduce/avoid false alarms, under ITEWS, ocean-bottom pressure recorders are placed to cover the two tsunamigenic zones.

Modeling at ITEWS

For estimating the possible tsunami traveltime and run-up heights at various locations, ITEWS uses the TUNAMI-N2 model of Imamura (2006). The model has been customized for the Indian Ocean. The computation is based on earthquake parameters and assumes maximum slip at the fault plane. A large data base of pre-run models has been created at the computation center. The data base has the surge heights and traveltimes of about 1800 points on the Indian coast. This helps in providing early warnings.

Decision support system at ITEWS

To pick up the closest scenario for an earthquake, a dedicated decision support system (DSS) has been developed at ITEWS (Fig. 5.3). This helps in assessing tsunami generation, the amplitude and the time of arrival at various locations in the Indian Ocean.

5.6 Assessment of a Tsunami Hazard on a Global Scale

Figure 5.4 (Løvholt et al. 2012) provides a global tsunami hazard overview. As can be noted from this figure, several areas in the vicinity of the Pacific seismic belt and areas in the vicinity of the Java-Sumatra seismic belt fall into the category where tsunamis exceeding heights of 5 m could occur. Løvholt et al. (2012) also pointed out areas for which adequate information is not available. In another

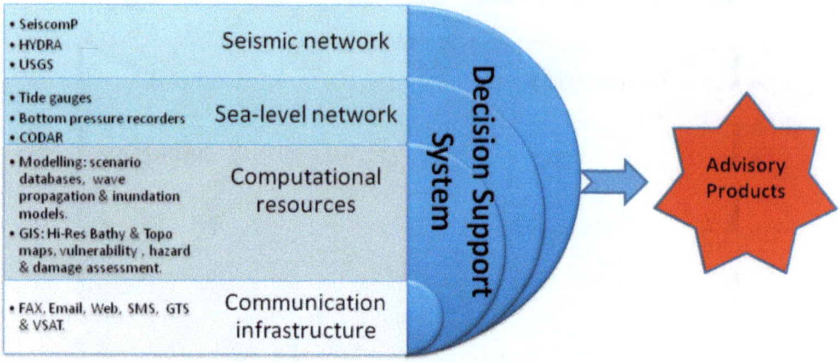

Fig. 5.3 Typical components of an end-to-end tsunami early warning system (after Nayak and Kumar 2011)

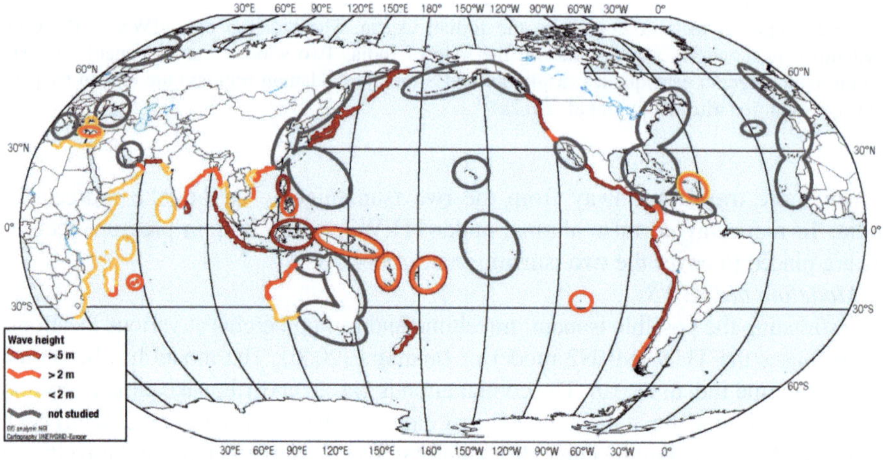

Fig. 5.4 The global tsunami scale hazard map. From the Global Assessment Report (UN-ISDR, 2009). After Løvholt et al. (2012)

interesting diagram (Fig. 5.5), the population living in areas that are tsunami prone in various countries is shown. Indonesia has more than 5 million people in tsunami prone areas followed by Japan. Another interesting diagram (Fig. 5.6) shows the absolute and relative GDP in tsunami prone areas.

5.7 Indian Tsunami Early Warning Center and the 11th April, 2012 Mw 8.5 Earthquake

An earthquake of Mw 8.5 occurred on April 11, 2012 at 08.38 Universal Time, with the epicenter at 2.40°N, 93.07°E and a focal depth of 10 km (Fig. 5.7). This was followed by another earthquake of Mw 8.2, with the epicenter at 0.87°N, 92.49°E

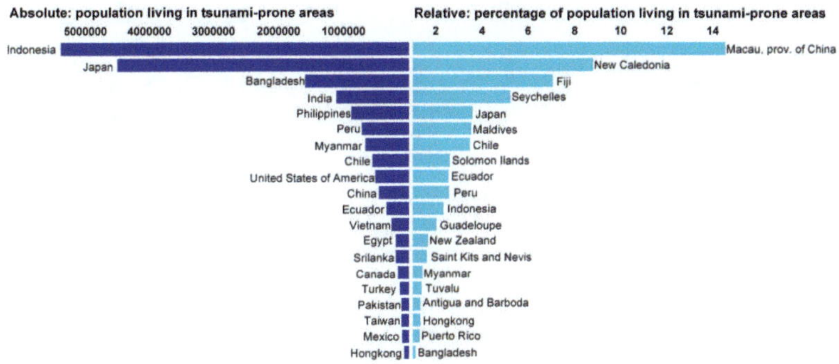

Fig. 5.5 The total and relative number of people exposed to tsunamis for a return period of about 475 years (after Løvholt et al. 2012). Here the risk in Bangladesh is shown as relatively high, however, new analysis of the tectonics and seismogenesis of the earthquakes in the region suggests a much lower risk to tsunamigenic earthquakes (Gupta and Gahalaut 2009 and Gahalaut et al. 2013)

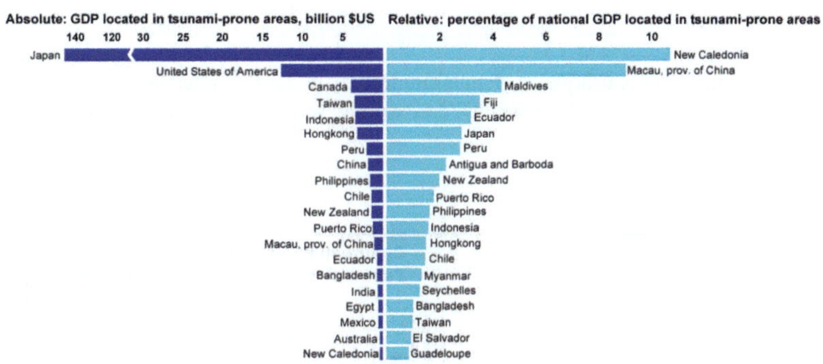

Fig. 5.6 The total and relative GDP exposed to tsunamis in some selected countries (after Løvholt et al. 2012)

on the same day at 10.43 Universal Time and also with a focal depth of 10 km. Both these earthquakes were located close to the epicenter of the devastating Mw 9.1 earthquake of 26th December 2004 (Fig. 3.1). ITEWS detected the first Mw 8.5 earthquake within 3 min 52 s of its occurrence and issued a necessary first advisory within 8 min from the origin time of the earthquake. The use of pre-run model simulations and the standard operation practice (SOP) at ITEWS placed only 3 zones in the Nicobar Islands under warning, necessitating movement of the coastal population to higher grounds (Kumar et al. 2012b). The Andaman Islands and the east coast of India were placed under 'Alert' status meaning threat and clearance of beaches only. These timely advisories avoided unnecessary panic and the evacuation of a large population as happened on March 28, 2005 after the Mw 8.5 Nias earthquake in the same region. The earthquakes generated a small ocean-wide tsunami that was no threat (Fig. 5.8). This clearly demonstrates the utility of ITEWS.

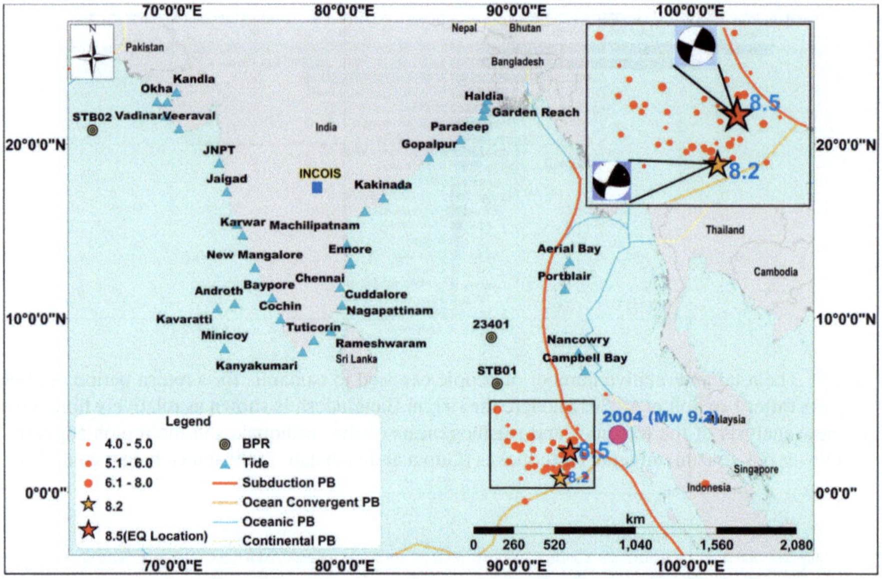

Fig. 5.7 Location of the earthquake of magnitude 8.5, its largest aftershock of magnitude 8.2 and other small aftershocks on 11 April 2012 as well as sea-level station locations (after Kumar et al. 2012b)

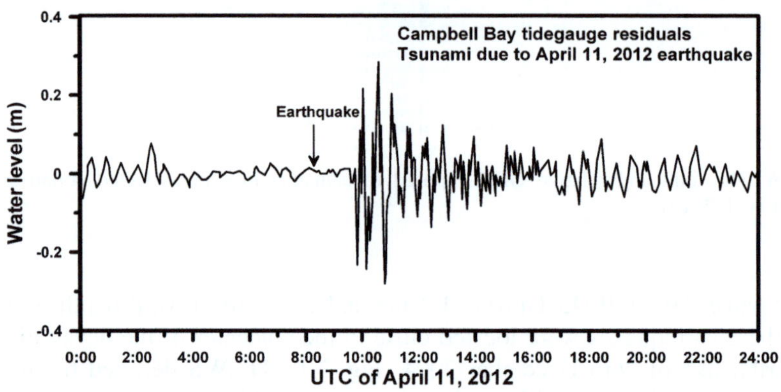

Fig. 5.8 Tsunami record at Campbell Bay (Great Nicobar island, India) tide gauge, about 500 km north of April 11, 2012 earthquake (after Nayak and Kumar 2011)

References

Gahalaut VK, Kundu B, Laishram SS, Catherine J, Kumar A, Singh MD, Tiwari RP, Chadha RK, Samanta SK, Ambikapathy A, Mahesh P, Bansal A, Narsaiah M (2013) Aseismic plate boundary in the Indo-Burmese wedge, northwest Sunda Arc Geology 41:235–238

Gica E, Spillane MC, Ttov VV, Chamberlin, CD, Newman JC (2008) Development of the forecast propagation database for NOAA's short term inundation forecast for tsunamis (SIFT), NOAA Technical Memorandum OAR PMEL 139. Contribution No. 2937 from NOAA/ Pacific Marine Environmental Laboratory

Greenslade DJM, Titov VV (2008) A comparison study of two numerical tsunami forecasting systems. Pure Appl Geophys 165:1991–2001

Gupta H, Gahalaut V (2009) Is the northern Bay of Bengal tsunamigenic? Bull Seismol Soc Am 99(6):3496–3501

Imamura F, Yalciner AC, Ozyurt G (1995) Tsunami modelling manual (TSUNAMI model). April http://www.tsunami.civil.tohoku.ac.jp/hokusai3/E/projects/manual-ver-3.1.pdf

Imamura F (2006) TUNAMI-N2 (Tohoku University's Numerical Analysis Model for Investigation of Near-field Tsunamis, version 2). Manual draft http://www.tsunami.civil.tohoku.ac.jp/hokusai3/J/projects/manual-ver-3.1.pdf. Accessed in 2006

Srinivasa Kumar K, Nayak S, Patanjali Kumar CH, Yadav RBS, Ajay Kumar B, Sunanda MV, Uma Devi E, Kiran Kumar N, Kishore SA, Shenoi SSC (2012a) Successful monitoring of the 11 April 2012 tsunami off the coast of Sumatra by India Tsunami early warning centre. Curr Sci 102:1519–1526

Srinivasa Kumar K, Nayak S, Patanjali Kumar CH, Yadav RBS, Ajay Kumar B, Sunanda MV, Uma Devi E, Shenoi SSC (2012b) Performance of the tsunami forecast system for the Indian Ocean. Curr Sci 102:110–114

Løvholt F, Glimsdal S, Harbitz CB, Zamora N, Nadim F, Peduzzi P, Dao H, Smebye H (2012) Tsunami hazard and exposure on the global scale. Earth Sci Rev 110:58–73

Nayak S, Srinivasa Kumar T (2008a) Indian tsunami warning system. In: International archives of the photogrammetry, remote sensing spatial information sciences, Part B4, Beijing, XXXVII, pp 1501–1506

Nayak S, Srinivasa Kumar T (2008b) The first tsunami warning centre in the Indian ocean. In: Risk wise, Tudor Rose Publishers, UK, pp 175–177

Nayak S, Srinivasa Kumar T (2008c) Addressing the risk of the tsunami in the Indian Ocean. J. South Asia Disaster Stud. 1(1):45–57

Nayak S, Kumar TS (2011) Tsunami watch and warning centres. In: Gupta H (ed) Encyclopedia of solid earth geophysics, vol 2. Springer, Dordrecht, pp 1498–1505

Titov VV, Gonzalez FI (1997) Implementation and testing of the method of splitting tsunami (MOST) model. NOAA Technical Memorandum ERL PMEL 112. Contribution No. 1927 from NOAA/Pacific Marine Environmental Laboratory

UNESCO-IOC, Manual and Guides (2006) Manual on sea level measurements and interpretation, vol IV, JCOMM Technical report no. 31, WMO/TD. No. 1339: An update to 2006. Paris, Intergovernmental Oceanographic Commission of UNESCO, p 78

Chapter 6
Where Are We?

Abstract Since the installation of the World Wide Standard Seismograph Network (WWSSN) and formulation of the 'Plate Tectonics Hypothesis' in the 1960s, there has been phenomenal progress in seismology. In relation to other science disciplines, the science of earthquakes and tsunamis is relatively young. Nevertheless, with the installation of several tsunami warning systems and expansion of seismological networks, our understanding of tsunami generation and advisories provided globally is significantly better now and improving.

A lot has happened since the occurrence of the 1755 Lisbon tsunami. At that time the debate was on whether an earthquake and a tsunami were due to the wrath of God or whether they were natural phenomena. The questionnaire that was sent to collect information regarding the felt and related observations of the public are relevant even today. It is said that the occurrence of the 1755 Lisbon earthquake and the investigations that followed laid the foundation of seismology. The Lisbon tsunami claimed an estimated one hundred thousand human lives. From Lisbon we move to the 2004 Mw 9.1 Sumatra–Andaman earthquake and the resultant tsunami that claimed an estimated two hundred and thirty thousand lives. By this time the science of earthquakes and tsunamis had been well established but the Indian Ocean region had relatively very few tsunamis: only three in the entire twentieth century and none of them being devastating. People were ignorant. The tsunami occurred on a Sunday morning and curiosity claimed several lives. Another important factor for the loss of the large number of human lives was the total disregard to the coastal region laws by several countries in the region that did not permit creation of infrastructure and residential properties within a stipulated distance from the high-tide line. For several countries the distance is 500 m. However, this is very often flouted in the South and Southeast Asian countries. Another factor was the holiday season when numerous tourists come to the balmy beaches of these countries. Also, an earthquake with Mw 9.1 was not expected in the region. The hypocenter was very close to heavily populated areas and there was not adequate time to respond. Moreover, the general public lacked training in defense against a tsunami. Then we had another unexpectedly large magnitude earthquake of Mw 9.0 on 11th March, 2011 in the coastal region of Japan, even though Japan is the most advanced country in the world as far as tsunami

H. K. Gupta and V. K. Gahalaut, *Three Great Tsunamis: Lisbon (1755), Sumatra–Andaman (2004) and Japan (2011)*, SpringerBriefs in Earth Sciences, DOI: 10.1007/978-94-007-6576-4_6, © Springer Science+Business Media Dordrecht 2013

related research and deployment of defensive measures are concerned. The main problem was the initial underestimation of the size of the earthquake. Moreover, the tsunami walls were constructed to tackle a tsunami generated by an Mw 8 earthquake and were not tall and strong enough to handle a tsunami generated by an Mw 9 earthquake. To top all these problems was the proximity of the nuclear power plants that were damaged. It is widely accepted that but for the defensive measures undertaken by Japan, the number of human lives lost would have been much greater.

Although the science of seismology has been practiced for over 250 years, the fundamental concept of plate tectonics evolved only in the late 1960s and early 1970s. The establishment of the World Wide Standard Seismograph Network of 100+ similar seismic stations globally in the years 1963/1964 provided an opportunity of systematic global coverage of earthquake occurrence. This contributed significantly to the development of the 'Plate Tectonics Hypothesis'. All the calculations of strain accumulation and earthquake occurrence are based on the 'Plate Tectonics Hypothesis' and therefore, it is appropriate to say that rigorous work on earthquake size and location is only about 50 years old. However, in these 50 years, considerable work has been done and we believe that we understand the phenomenon of earthquake occurrence and tsunami generation a lot better now.

As far as tsunamis are concerned, significant ground has been covered in the last 8 years. One of us (HKG) very distinctly remembers that on 26th December, 2004 when the tsunami hit the Andaman and Nicobar group of islands, most of the media people in India did not know the word 'Tsunami'. In the years to follow India succeeded in setting up the 'Indian Tsunami Early Warning System (ITEWS)' with state of the art tsunami watch and warning capabilities. The two other systems that were established following the 2004 Sumatra–Andaman tsunami are the 'German-Indonesian Tsunami Early Warning System (GITEWS)' and the 'Joint Australia Tsunami Warning Center (JATWC)'. On 28th March, 2005 the Nias Mw 8.7 earthquake occurred late in the evening (9:39 pm, Indian Standard Time) and a tsunami warning was issued that caused massive evacuation on the east coast of India. No tsunami requiring evacuation occurred in the region. This caused immense inconvenience to a huge population on the east coast of India. Let us compare that situation with what happened on 11th April 2012 when two earthquakes of Mw 8.6 and 8.2 occurred within 2 h of one another. These were located close to the epicenter of the December 26, 2004 Sumatra–Andaman earthquake of Mw 9.1. The timely advisories by ITEWS, as discussed in Chap. 5, did not create any panic. Moreover, within 30 min of the occurrence of both these earthquakes, it was discovered that these were 'strike-slip' motion (Fig. 1.1) dominated earthquakes, which are not conducive for the production of a tsunami.

In the last decade, new improved methods have been developed to understand tsunami propagation. As the majority of tsunamis are caused by earthquakes, the earthquake monitoring networks based on seismological and geodetic methods, have expanded tremendously, which are not only continuously providing information about earthquake occurrence but also about the processes causing them. It is important to communicate these results to the public so that they can incorporate them into their construction activities. The laws governing construction activities need to evolve continuously and should be implemented in a most strict manner. Effective and timely warnings about tsunamis and earthquakes and their communication to the local public can help in reducing the loss of property and lives. This will ultimately lead to a better and objective assessment of seismic and tsunami hazards so that tomorrow's world will be safer.

MIX
Papier aus verantwortungsvollen Quellen
Paper from responsible sources
FSC® C105338

If you have any concerns about our products,
you can contact us on
ProductSafety@springernature.com

In case Publisher is established outside the EU,
the EU authorized representative is:
Springer Nature Customer Service Center GmbH
Europaplatz 3, 69115 Heidelberg, Germany

Printed by Libri Plureos GmbH
in Hamburg, Germany